QUADRUPEDS OF NORTH AMERICA

THE

QUADRUPEDS

OF

NORTH AMERICA

BY

JOHN JAMES AUDUBON, F R. S., &c. &c.

AND

THE REV. JOHN BACHMAN, D. D., &c. &c.

VOL. III.

NEW-YORK:
PUBLISHED BY V. G. AUDUBON.
1854.

R. CRAIGHEAD, PRINTER AND STEREOTYPER.
53 *Vesey Street, New York*

TABLE OF CONTENTS.

TABLE OF CONTENTS.

Plate CI.

Drawn from Nature by J.W. Audubon.

On Stone by W.E. Hitchcock

The Jaguar.

Lith.ᵈ Printed & Col.ᵈ by J.T. Bowen, Philadᵃ.

QUADRUPEDS OF NORTH AMERICA.

————————

FELIS ONCA.—LINN.

JAGUAR.

PLATE CI.—FEMALE.

F. Supra fulva, subtus albus ; corpore ocellis annularibus nigris ornato, in series subparallelis per longitudinem dispositis ; ocellis, punctis nigris subcentralibus, in signitis.

CHARACTERS.

Yellow, with a white belly ; body marked with open black circle-like figures, each containing one or more nearly central black dots ; these black, circle-like markings disposed in nearly longitudinal parallel lines.

SYNONYMES.

FELIS ONCA. Linn. Syst. Natur. vol. xii. p. 61 ; Gmel. vol. i. p. 77, pl. 4 (4 ed.).
   "   "   Schreber, Säugth. p. 388, pl. 6.
   "   "   Erxleben Syst. p. 513, pl. 9.
   "   "   Zimm. Geogr. Gesch. ii. pp. 162, 268.
   "   "   Cuv. Ann. du Mus. xiv. p. 144. 4 T. 16.
   "   "   " Regne Animale, vol. i. p. 260. Ossements Fossiles, vol. iv. p. 417.
   "   "   F. Cuv. Dict. Sci. Nat., vol. viii. p. 223.
   "   "   Desm. in Nouv. Dict., vol. vi. p. 97, pl. 4.
   "   "   "   Mammal., pp. 219, 338.
   "   "   Desmoulins, Dict. Class 3d, p. 498.
   "   "   Temm. Monog., p. 136.
   " PANTHERA. Schreber, t, 99.
   " CAUDA ELONGATA. Brown's Jamaica.
TIGRIS REGIA. Briss. Regne Animale, p. 269, fig. 7.

TLATLAUHQUI OCELOTL. TIGRIS MEXICANA. Hernandez, Mex., p. 498, fig. c.
JAGUARA. Marcgr. Brazil, p. 235, fig. c.
JAGUAR. Buff. Nat. Hist., tom. ix. p. 201.
YAGOUARÉTÉ. D'Azara, vol. i. p. 114.
BRAZILIAN PANTHER. Pennant's Synopsis, pp. 127, 176.
     "   TIGER. Pennant's Quadrupeds, p. 286.
ONZA PINTADO. *Lusitanis,* in Bresil. Cumang *Maconis.*
FELIS JAGUAR. Hamilton Smith. Griffith's An. Kingdom, vol. v. p. 164.
     "   ONCA. Harlan, Fauna, p. 95.

DESCRIPTION.

The Jaguar compares with the Asiatic tiger in size and in shape; its legs, however, are shorter than those of the royal tiger, although its body is perhaps as heavy.

Head, large; jaws, capable of great expansion; incisors, large, and slightly curved inwards; ears, rather small, rounded, clothed with short hairs on the inside. Body, rather inclining to be stout, and shorter and less elegant than the cougar: at the shoulders the Jaguar is not much more raised from the earth, but it stands higher from the ground near the rump.

Feet, clothed with hair covering the retractile nails; the pads of the feet, naked; a few hairs between the toes; tail, long, and generally half elevated when walking; whiskers, few, strong, and bristly.

Hair of two kinds; the longest (which is only from four to five eighths of an inch in length) is the coarser; the shortest is a softer and finer fur, and is not very thickly distributed.

COLOUR.

Where the black markings do not prevail, the hairs are light greyish-brown at the roots and on the surface rich straw-yellow, deepest near the shoulders and back, and paler on the sides and legs; nose to near the eye nearly a uniform lightish-brown; forehead spotted with black in somewhat curved lines, the spots becoming larger towards the back of the head; whiskers black at the roots, then white for two thirds of their length to the points; lips and chin, white; a black line on the sides of the mouth; around the eye, whitish-yellow; iris, light-yellow; a black stripe between the ears on the back part of the head. There is no white patch behind the ear, as in the cougar and the wild cat.

All the black spots on the body are composed of hairs which are black from their roots; outer edge of the ear, black for half an inch in width;

a row of black spots running along the back to and beyond the root of the tail for about a foot along its upper surface ; the sides of the body are marked with black rings of irregular and somewhat oval shapes, with yellow-brown centres having dots of pure black in them. These black rings are, on the edge of the back somewhat diamond shaped, with from one to three little black spots inside. Many of these circles or squares are not perfect : some are formed by several dots and curved black patches which turn inwards.

On the shoulders and the outer surfaces of the legs, these rings or squares are succeeded by black spots or patches lessening in size as they approach the claws. The hair on the under surface is dull-white from the roots, with large patches of black ; belly, inner sides of legs, and throat, white, blotched or spotted with black. These patches are irregular in size, being from one eighth of an inch to two inches in extent. Tail, general colour spotted black on a yellow ground, like the outsides of the legs.

A living Jaguar from Mexico which we examined in its cage at Charleston, became very beautiful after shedding its hair in spring : the general colour of its body was bright-yellow, and the rings and spots were brilliant black.

There was another living specimen in the same collection, from Brazil, which resembled the one from Mexico in its general markings, but was larger, more clumsy, and had shorter and thicker legs. There were, however, no characters by which the species could be separated.

DIMENSIONS.

	Feet.	Inches.
From point of nose to root of tail, - - - -	4	1
Length of tail, - - - - - - -	2	1 (?)
Height of ear, - - - - - - -		2¾
Shoulder to end of claw, - - - - - -	2	
Length of largest claw, - - - - - -		2
Around the wrist, - - - - - - -		7½
" " chest, - - - - - -	3	
" " head, - - - - - -	1	9¾
Breadth between the eyes, - - - - - -		3

HABITS.

Alike beautiful and ferocious, the Jaguar is of all American animals unquestionably the most to be dreaded, on account of its combined strength, activity, and courage, which not only give it a vast physical power over other wild creatures, but enable it frequently to destroy man.

Compared with this formidable beast, the cougar need hardly be dreaded more than the wild cat; and the grizzly bear, although often quite as ready to attack man, is inferior in swiftness and stealthy cunning. To the so much feared tiger of the East he is equal in fierceness; and it is owing, perhaps, to his being nocturnal in his habits to a great extent, that he seldom issues from the deep swamps or the almost impenetrable thickets or jungles of thorny shrubs, vines, and tangled vegetation which compose the chaparals of Texas and Mexico, or the dense and untracked forests of Central and Southern America, to attack man. From his haunts in such nearly unapproachable localities, the Jaguar roams forth towards the close of the day, and during the hours of darkness seizes on his prey. During the whole night he is abroad, but is most frequently met with in moonlight and fine nights, disliking dark and rainy weather, although at the promptings of hunger he will draw near the camp of the traveller, or seek the almost wild horses or cattle of the ranchero even during daylight, with the coolest audacity.

The Jaguar has the cunning to resort to salt-licks, or the watering-places of the mustangs and other wild animals, where, concealing himself behind a bush, or mounting on to a low or sloping tree, he lies in wait until a favorable opportunity presents itself for springing on his prey. Like the cougar and the wild cat, he seeks for the peccary, the skunk, opossum, and the smaller rodentia; but is fond of attacking the larger quadrupeds, giving the preference to mustangs or horses, mules, or cattle. The colts and calves especially afford him an easy prey, and form a most important item in the grand result of his predatory expeditions.

Like the lion and tiger, he accomplishes by stealth or stratagem what could not be effected by his swiftness of foot, and does not, like the untiring wolf, pursue his prey with indomitable perseverance at top speed for hours together, although he will sneak after a man or any other prey for half a day at a time, or hang on the skirts of a party for a considerable period, watching for an opportunity of springing upon some person or animal in the train.

Col. HAYS and several other officers of the Rangers, at the time J. W. AUDUBON was at San Antonio de Bexar, in 1845, informed him that the Jaguar was most frequently found about the watering-places of the mustangs, or wild horses, and deer. It has been seen to spring upon the former, and from time to time kills one; but it is much more in the habit of attacking colts about six months old, which it masters with great ease. Col. HAYS had killed four Jaguars during his stay in Texas. These animals are known in that country by the Americans as the "Leopard," and by the Mexicans as the "Mexican tiger." When lying in wait at or near the watering-places of deer or horses, this savage beast exhibits great patience and perseverance, remaining for hours crouched down, with head depressed, and still as death. But when some luckless animal approaches, its eyes seem to dilate, its hair bristles up, its tail is gently waved backwards and forwards, and all its powerful limbs appear to quiver with excitement. The unsuspecting creature draws near the dangerous spot; suddenly, with a tremendous leap, the Jaguar pounces on him, and with the fury of an incarnate fiend fastens upon his neck with his terrible teeth, whilst his formidable claws are struck deep into his back and flanks. The poor victim writhes and plunges with fright and pain, and makes violent efforts to shake off the foe, but in a few moments is unable longer to struggle, and yields with a last despairing cry to his fate. The Jaguar begins to devour him while yet alive, and growls and roars over his prey until his hunger is appeased. When he has finished his meal, he sometimes covers the remains of the carcass with sticks, grass, weeds, or earth, if not disturbed, so as to conceal it from other predacious animals and vultures, until he is ready for another banquet. The Jaguar often lies down to guard his prey, after devouring as much as he can. On one occasion a small party of Rangers came across one while feeding upon a mustang. The animal was surrounded by eight or ten hungry wolves, which dared not interfere or approach too near "the presence." The Rangers gave chase to the Jaguar, on which the wolves set up a howl or cry like a pack of hounds, and joined in the hunt, which ended before they had gone many yards, the Jaguar being shot down as he ran, upon which the wolves went back to the carcass of the horse and finished him.

The Jaguar has been known to follow a man for a long time. Colonel HAYS, whilst alone on a scouting expedition, was followed by one of these animals for a considerable distance. The colonel, who was aware that his footsteps were scented by the animal, having observed him on his trail a little in his rear, had proceeded a good way, and thought that the Jaguar had left, when, having entered a thicker part of the wood, he heard a stick crack, and being in an Indian country, "whirled round," expecting to face a Wakoe; but instead of a red-skin, he saw the Jaguar, about half-crouched, looking "right in his eye," and gently waving his tail. The colonel, although he wished not to discharge his gun, being in the neighborhood of Indians who might hear the report, now thought it high time to shoot, so he fired, and killed him in his tracks. "The skin," as he informed us, "was so beautiful, it was a pleasure to look at it."

These skins are very highly prized by the Mexicans, and also by the Rangers; they are used for holster coverings and as saddle cloths, and form a superb addition to the caparison of a beautiful horse, the most important animal to the occupants of the prairies of Texas, and upon which they always show to the best advantage.

In a conversation with General HOUSTON at Washington city, he informed us that he had found the Jaguar east of the San Jacinto river, and abundantly on the head waters of some of the eastern tributaries of the Rio Grande, the Guadaloupe, &c.

These animals, said the general, are sometimes found associated to the number of two or more together, when they easily destroy horses and other large quadrupeds. On the head waters of the San Marco, one night, the general's people were aroused by the snorting of their horses, but on advancing into the space around could see nothing, owing to the great darkness. The horses having become quiet, the men returned to camp and lay down to rest as usual, but in the morning one of the horses was found to have been killed and eaten up entirely, except the skeleton. The horses on this occasion were hobbled and picketed; but the general thinks the Jaguar frequently catches and destroys wild ones, as well as cattle. The celebrated BOWIE caught a splendid mustang horse, on the rump of which were two extensive scars made by the claws of a Jaguar or cougar. Such instances, indeed, are not very rare.

Capt. J. P. McCOWN, U. S. A., related the following anecdote to us :— At a camp near the Rio Grande, one night, in the thick, low, level musquit country, when on an expedition after Indians, the captain had killed a beef which was brought into camp from some distance. A fire was made, part of the beef hanging on a tree near it. The horses were picketed around, the men outside forming a circular guard. After some hours of the night had passed, the captain was aroused by the soldier next him saying, "Captain, may I shoot?" and raising himself on his arm, saw a Jaguar close to the fire, between him and the beef, and near it, with one fore-foot raised, as if disturbed; it turned its head towards the captain as he ordered the soldier not to fire, lest he should hurt some one on the other side of the camp, and then, seeming to know it was discovered, but without exhibiting any sign of fear, slowly, and with the stealthy, noiseless pace and attitude of a common cat, sneaked off.

The Jaguar, in its South American range, was long since noticed for its ferocity by HUMBOLDT and others. In some remarks on the American animals of the genus felis, which we find in the Memoirs of the Wernerian Nat. Hist. Society of Edinburgh, vol. iv., part 2, p. 470, it is stated that the Jaguar, like the royal tiger of Asia, does not fly from man when it is dared to close combat, when it is not alarmed by the great number of its assailants. The writer quotes an instance in which one of these animals had seized a horse belonging to a farm in the province of Cumana, and dragged it to a considerable distance. " The groans of the dying horse," says HUMBOLDT, " awoke the slaves of the farm, who went out armed with lances and cutlasses. The animal continued on its prey, awaited their approach with firmness, and fell only after a long and obstinate resistance." In the same article, the writer states that the Jaguar leaps into the water to attack the Indians in their canoes on the Oronoko. This animal called the Yagouarété in Paraguay if we are not mistaken, the foregoing article goes on to say, is described by gentlemen who have hunted it in that country, as a very courageous and powerful animal, of great activity, and highly dangerous when at bay. He also says : " Both this species and the puma are rendered more formidable by the facility with which they can ascend trees.

" A very beautiful Jaguar from Paraguay was some time ago carried alive to Liverpool. When the animal arrived, it was in full health, and though not fully grown was of a very formidable size and strength. The captain who brought it could venture to play with it, as it lay on one of the boats on deck, to which it was chained ; but it had been familiarized to him from the time it was the size of a small dog."

In Griffith's Cuvier, vol. ii. p. 457, it is stated in a quotation from D'Azara, that the Jaguar is reported to " stand in the water out of the stream, and drop its saliva, which, floating on the surface, draws the fish after it within reach, when it seizes them with the paw, and throws them ashore for food." At the same page, it is said, " The Jaguar is hunted with a number of dogs, which, although they have no chance of destroying it themselves, drive the animal into a tree, provided it can find one a little inclining, or else into some hole. In the first case the hunters kill it with fire-arms or lances ; and in the second, some of the natives are occasionally found hardy enough to approach it with the left arm covered with a sheep-skin, and to spear it with the other—a temerity which is frequently followed with fatal consequences to the hunter."

The Jaguars we examined in a menagerie at Charleston had periodical fits of bad temper : one of them severely bit his keeper, and was ready to give battle either to the Asiatic tiger or the lion, which were kept in separate cages.

We add some extracts, with which we hope our readers will be interested :

" In the province of Tucuman, the common mode of killing the Jaguar is to trace him to his lair by the wool left on the bushes, if he has carried

off a sheep, or by means of a dog trained for the purpose. On finding the
enemy, the gaucho puts himself into a position for receiving him on the
point of a bayonet or spear at the first spring which he makes, and thus
waits until the dogs drive him out—an exploit which he performs with
such coolness and dexterity that there is scarcely an instance of failure.
In a recent instance related by our capitaz, the business was not so quickly
completed. The animal lay stretched at full length on the ground, like a
gorged cat. Instead of showing anger and attacking his enemies with
fury, he was playful, and disposed rather to parley with the dogs with
good humour than to take their attack in sober earnestness. He was now
fired upon, and a ball lodged in his shoulders, on which he sprang so
quickly on his watching assailant that he not only buried the bayonet in
his body, but tumbled over the capitaz who held it, and they floundered on
the ground together, the man being completely in his clutches. 'I
thought,' said the brave fellow, 'I was no longer a capitaz, while I held
my arm up to protect my throat, which the animal seemed in the act of
seizing; but when I expected to feel his fangs in my flesh, the green fire
of his eyes which blazed upon me flashed out in a moment. He fell on me,
and expired at the very instant I thought myself lost for ever.'"—*Captain
Andrews's Travels in South America*, vol. i. p. 219.

"Two Indian children, a boy and girl eight or nine years of age, were
sitting among the grass near the village of Atures, in the midst of a
savannah. It was two in the afternoon when a Jaguar issued from the
forest and approached the children, gambolling around them, sometimes
concealing himself among the long grass, and again springing forward, with
his back curved and his head lowered, as is usual with our cats. The
little boy was unaware of the danger in which he was placed, and became
sensible of it only when the Jaguar struck him on the head with one of his
paws. The blows thus inflicted were at first slight, but gradually became
ruder. The claws of the Jaguar wounded the child, and blood flowed
with violence. The little girl then took up a branch of a tree, and struck
the animal, which fled before her. The Indians, hearing the cries of the
children, ran up and saw the Jaguar, which bounded off without showing
any disposition to defend itself."—*Humboldt's Travels and Researches, &c.*,
Edinburgh, 1833, p. 245.

HUMBOLDT speculates on this cat-like treatment of the children, and we
think it very likely that occasionally the Jaguar plays in a similar manner
with its prey, although we have not witnessed it, nor heard of any
authentic case of the kind.

D'AZARA says (vol. i. p. 116) that the black Jaguar is so rare that in
forty years only two had been killed on the head waters of the river

Parana. The man who killed one of these assured him that it did not differ from the Jaguar (Yagouareté), except that it was black, marked with still blacker spots, like those of the common Jaguar.

The Jaguar generally goes singly, but is sometimes accompanied by his favourite female. The latter brings forth two young at a time, the hair of which is rougher and not so beautiful as in the adult. She guides them as soon as they are able to follow, and supplies and protects them, not hesitating to encounter any danger in their defence.

The Jaguar, according to D'AZARA, can easily drag away a horse or an ox ; and should another be fastened or yoked to the one he kills, the powerful beast drags both off together, notwithstanding the resistance of the terrified living one. He does not conceal the residue of his prey after feeding : this may be because of the abundance of animals in his South American haunts. He hunts in the stealthy manner of a cat after a rat, and his leap upon his prey is a very sudden, quick spring : he does not move rapidly when retreating or running. It is said that if he finds a party of sleeping travellers at night, he advances into their midst, and first kills the dog, if there is one, next the negro, and then the Indian, only attacking the Spaniard after he has made this selection ; but generally he seizes the dog and the meat, even when the latter is broiling on the fire, without injuring the men, unless he is attacked or is remarkably hungry, or unless he has been accustomed to eat human flesh, in which case he prefers it to every other kind. D'AZARA says very coolly, "Since I have been here the Yagouareté (Jaguars) have eaten six men, two of whom were seized by them whilst warming themselves by a fire." If a small party of men or a herd of animals pass within gunshot of a Jaguar, the beast attacks the last one of them with a loud roar.

During the night, and especially in the love season, he frequently roars, uttering in a continued manner, *pou, pou, pou*.

It is said that when the Spaniards settled the country from Montevideo to Santa-Fé de Vera Cruz, so many Jaguars were found that two thousand were killed annually, but their numbers have been greatly diminished (D'AZARA, vol. i. p. 124). We have no positive information as to the present average annually killed, but presume it not to exceed one tenth the above number.

GEOGRAPHICAL DISTRIBUTION.

This species is known to exist in Texas, and in a few localities is not very rare, although it is far from being abundant throughout the state. It is found on the head waters of the Rio Grande, and also on the Nueces.

Towards the west and southwest it extends to the mountainous country beyond El Paso. HARLAN speaks of its being occasionally seen east of the Mississippi. This we think somewhat doubtful. It inhabits Mexico and is frequently met with in almost every part of Central America. HUMBOLDT mentions having heard its constant nightly screams on the banks of the Oronoco. It is known to inhabit Paraguay and the Brazils, and may be regarded as the tiger of all the warmer parts of America, producing nearly as much terror in the minds of the feeble natives as does its congener, the royal tiger, in the East. It is not found in Oregon, and we have not met with any account of it as existing in California.

GENERAL REMARKS.

BUFFON, in describing the habits of the Jaguar, appears to have received his accounts of the timidity of this species from those who referred to the Ocelot, which is generally admitted to be a timid animal. He erroneously supposed that when full grown it did not exceed the size of an ordinary dog, in which he egregiously underrated its dimensions. It is certainly a third heavier than the Cougar, and is not only a more powerful, but a far more ferocious animal. This species exhibits some varieties, one of which, the black Jaguar, is so peculiar that it has been conjectured that it might be entitled to a distinct specific name. The exceeding rarity, however, of the animal, and the variations to which nearly all the species of this genus are subject, induce us to set it down as merely a variety. It must be observed that it is rare to find two specimens of uniform colour ; indeed the markings on each side of the same animal are seldom alike. BUFFON (vol. v. p. 196, pl. 117–119) has given three figures of the Jaguar, the first and third of which we consider as the Ocelot, and the second as probably the Panther (*F. Pardus*) of the eastern continent. HAMILTON SMITH, in GRIFFITH'S CUVIER (vol. ii. pp. 455, 456), has given us two figures of this species, differing considerably in colour and markings : the former is very characteristic. He has named this species *Felis Jaguar*, which is inadmissible. There is some resemblance in this species to the panther (*F. Pardus*), as also to the leopard (*F. Leopardus*) of Africa, but they are now so well described as distinct species that it is scarcely necessary to point out the distinctive marks of each. BUFFON's panthère femelle, pl. 12, and SHAW's, Gen. Zool., Part I., pl. 84, evidently are figures of our Jaguar.

On Stone by W.E. Hitchcock.

Large Tailed Skunk.

Drawn from Nature by J.W.Audubon   Lith.ᵈ Printed & Col.ᵈ by J.T.Bowen, Philad.ᵃ

MEPHITIS MACROURA.—Licht.

LARGE-TAILED SKUNK.

PLATE CII.—Male.

M. magnitudine felis cati (domestica), fusco-niger, striis duabus albis dorsalibus, vitta alba frontali, cauda capite longiore.

CHARACTERS.

Size of the domestic cat ; general colour, brownish-black ; a white stripe on each side of the back, and on the forehead ; tail longer than the head.

SYNONYMES.

MEPHITIS MACROURA. Licht., Darstellung neuer oder wenig bekannter Säugthiere, Berlin, 1827–34, Tafel xlvi.
" MEXICANUS GRAY. Loudon's Mag., p. 581. 1837.

DESCRIPTION.

Body, as in other species of this genus, stout ; head, small ; nose short, rather acute, and naked ; ears short, rounded, clothed with short hair on both surfaces ; eyes, small ; claws, slender and weak ; soles of the feet naked.

The body is covered with two kinds of hair ; the first long and glossy, the fur underneath soft and woolly ; tail very long, rather bushy, covered with long hairs, and without any of the softer and shorter fur.

COLOUR.

There are slight variations in the markings of the specimens we examined in the museums of Berlin and London, and in those we possess. This species appears, however, to be less eccentric in colour and markings than the common skunk *M. chinga.*

In the specimen from which our figure was made, there is a rather broad longitudinal white stripe running from the nose to near the back of the head ; upper surface of neck and back, white, with a narrow black dorsal stripe beginning on the middle of the back and running down on the upper surface of the tail ; a spot of white under the shoulder, and another along the flanks ; the hairs on the tail are irregularly mixed with white and black ; under surface black.

Another skin from the same region has a narrower stripe on the forehead, the usual white stripes from the back of the head along the sides nearly meeting again at the root of the tail, leaving the dorsal black patch very much broader than in the specimen just described, and of an oval shape ; the tail contains a greater number of black hairs, and towards the tip is altogether black ; sides, legs, and whole under surface, black.

LICHTENSTEIN's figure resembles this specimen in form and markings, with the exceptions that it represents scarcely any black patch on the back, and that it exhibits a longitudinal white stripe running from the shoulder to the hip. LICHTENSTEIN has also described and figured the young of this species, which very closely resembles the adult.

DIMENSIONS.

Male.—Killed January 28, 1846.

	Feet.	inches.
From point of nose to root of tail, - - -	1	4
Tail (vertebræ), - - - - - -	1	1
" to end of hair, - - - - -	1	6
Between ears, - - - - - - -		2¼
Girth around the body, behind fore-legs, - -		9
" " belly, - - - -	1	2¼
Height from sole of fore-foot to top of shoulders,		8¾

Weight, 4½lb.—specimen fat.

HABITS.

In Texas, during the winter of 1845–6, specimens of this skunk were obtained by J. W. AUDUBON ; the first he met with was seen on one of the high and dry prairies west of Houston, on the road to Lagrange ; this was, however, only a young one. It was easily caught, as these animals never attempt to escape by flight, depending on the fetid discharges which they, like the common skunk, eject, to disgust their assailant and cause him to leave them in safety. By throwing sticks and clods of dirt at this young one, he was induced to display his powers in this way, and teased until he had emptied the glandular sacs which contain the detestable secretion. He was then comparatively disarmed, and by thrusting a forked stick over the back of his head, was pinned to the ground, then seized and thrust into a bag, the mouth of which being tied up, he was considered safely captured, and was slung to one of the pack-saddles of the baggage-mules. The fetor of this young skunk was not so horrid as that of the common species (*Mephitis chinga*).

On arriving at the camping ground for the night, the party found that their prisoner had escaped by gnawing a hole in the bag, being unobserved by any one.

This species is described as very common in some parts of Texas, and its superb tail is now and then used by the country folks by way of plume or feather in their hats. J. W. AUDUBON, in his Journal, remarks : " We were much amused at the disposition manifested by some of the privates in the corps of Rangers, to put on extra finery when opportunity offered. At one time a party returned from a chase after Indians whom they had over-taken and routed. Several of them had whole turkey-cocks' tails stuck on one side of their hats, and had long pendant trains of feathers hanging behind their backs, which they had taken from the ' braves' of the Wakoes. One young fellow, about eighteen years of age, had a superb head-dress and suit to match, which he had taken from an Indian, whom, to use his own expression, he had scared out of it : he had, to complete the triumphal decoration of his handsome person, painted his face all the colours of the rainbow, and looked fierce enough. In contrast with these freaks of some of the men, we noticed that their tried and chivalrous leaders, HAYS, WALKER, GILLESPIE, and CHEVALIER, were always dressed in the plainest costume the ' regulations' permitted."

The Large-Tailed Skunk feeds upon snakes, lizards, insects, birds' eggs, and small animals ; and it is said that at the season when the pecan (*Carya olivaeformis*) ripens, they eat those nuts, as well as acorns. This is strange, considering their carnivorous formation. They burrow in winter, and live in hollows and under roots. They produce five or six young at a birth.

We are indebted to Col. GEO. A. McCALL, U. S. A., for the following interesting account of an adventure with one of these Skunks, which, besides being written in an entertaining and lively manner, sets forth in a strong light the dread the very idea of being defiled by these offensive brutes causes in every one who has ever been in those parts of the country they inhabit :—

" In New Mexico, in September last, returning from Los Vegas to Santa Fé, I halted for the night at Cottonwood creek. Here, I pitched my tent on the edge of a beautiful grove of the trees (*Populus angulatus*) which give name to the stream.

" Wishing to reach my destination at an early hour on the morrow, I directed the men to be up before day, in order that they might feed their horses, get their breakfast, and be ready to take the road as soon as it was fairly daylight. After a refreshing sleep, I awoke about an hour before day, and the familiar sound of my horse munching his corn by the side of my tent, where he was usually picketed, informed me that my men were already astir. At this hour, the moon, almost at the full, was low in the west, and flung its mellow light adown the mountain gorge, in rays that were nearly horizontal. And therefore, not finding it necessary to strike a light, I was on the point of rising, when I heard, as I thought, my servant opening the mess-basket, which stood near the foot of my bed. I spoke to him ; but receiving no answer, I turned my eyes in that direction, and discovered on the front wall of my tent a little shadow playing fantastically over the canvas, upon which the moon's rays fell, after passing over my head. With a hunter's eye, I at once recognized in this shadow the outline of the uplifted tail of a *Mephitis Macroura*, vulgo Large-Tailed *Skunk*, whose body was concealed from my view behind the mess-basket. Into this, doubtless attracted by the scent of a cold boiled bacon-ham, he was evidently endeavouring to effect an entrance.

"Being well acquainted with his habits and character, I knew I must manage to get rid of my visitor without seriously alarming or provoking him, or I should in all probability be the sufferer. I therefore thought I would at first, merely in a quiet way, signify my presence ; on discovering which, perhaps, he would take the hint, and his departure at the same time. So, 'I coughed and cried hem!' but my gentleman only raised his head above the top of the basket for a moment, and then renewed his efforts to lift the lid. I now took up one of my boots that lay by my bed, and struck the heel smartly against the tent-pole. Again the intruder raised his head, and regarded me for a moment ; after which he left the basket and passed round the foot of my bed, which, I should mention, was spread upon the ground. At first, I thought he had, indeed, taken the hint, and was about to slope off. But I had, in fact, only excited his curiosity ; and the next moment, to my horror, I saw him turn up by the side of my bed, and come dancing along with a dainty, sidling motion, to examine into the cause of the noise. His broad white tail was elevated, and jauntily flirted from side to side as he approached. In fact, his approach was the sauciest and most provokingly deliberate thing conceivable. As every step brought him nearer to my face, the impulse I felt to bolt head-foremost through the opposite side of the tent, was almost irresistible ; but I well knew that any sudden motion on my part, whilst in such close proximity to the rascal, would be very apt so to startle him as to bring upon me that which I was seeking to escape, and of which I was, in truth, in mortal dread ; whilst, on the other hand, I was equally aware that my safety lay ın keeping perfectly still, for it was quite probable that the animal, after having satisfied his curiosity, would, if uninterrupted, quietly take his departure. The trial was a severe one, for the next moment the upright white tail was passing within a foot of my very face. I did not flinch, but kept my eye upon it, although the cold sweat broke out upon my forehead in great globules. At length the fellow finding nothing to alarm him, turned about and with a sidelong motion danced back again to the mess-basket. Finding now that he had no thought of taking himself away, I exclaimed internally, 'Mortal man cannot bear a repetition of what I have just experienced!' and laid my hand upon my rifle, which stood at my head. I weighed the chances of killing the animal so instantly dead that no discharge of odour would take place ; but just at this moment he succeeded in raising the top of the basket and I heard his descent among the spoons. 'Ha! ha! old fellow, I have you now!' I said to myself: and the next instant I was standing on the top of the mess-basket, whither I had got without the slightest noise, and where I now heard the rascal rummaging my things little suspecting that he was at the time a prisoner. I called my servant—a negro. George made his appearance, and as he opened the front of the tent paused in surprise at seeing me standing *en dishabille* on the top of the mess-basket. 'George,' said I, in a quiet tone, 'buckle the straps of this basket.' George looked still more surprised on receiving the order, but obeyed it in silence. I then stepped gently off, and said, 'Take this basket very carefully, and without shaking it, out yonder, in front, and set it down easily.' George looked still more bewildered ; but, accustomed to obey without question, did as he was directed. After he had carried the basket off to a considerable distance, and placed it on the ground, he looked back at the door of the tent, where I still stood, for further orders. 'Unbuckle the straps,' said I : it was done. 'Raise the top of the basket :' he did so ; while at the same time, elevating my voice, I continued, '*and let that d——d Skunk out !*' As the last words escaped from my lips the head and tail of the animal appeared in sight, and George, giving vent to a scream of surprise and fear, broke away like a quarter-horse, and did not stop until he had put a good fifty yards between himself and the mess-basket. Meanwhile, the Skunk, with the same deliberation that had marked his previous course (and which, by the way, is a remarkable trait in the character of this animal), descended the side of the basket, and, with tail erect, danced off in a direction down the creek, and finally disappeared in the bushes. I then, having recovered from a good fit of laughter, called to George, who rather reluctantly made his appearance before me. He was still a little out of breath, and with some agitation, thus delivered himself, 'Bless God, massa, if I had known there was a Skunk in the mess-basket, I never would have touched it *in this world !*' 'I knew that well enough, George, and that was the reason I did not tell you of it.'

"It is only necessary further to say that the animal, having been neither alarmed nor provoked in any way, did not on this occasion emit the slightest odour ; nor was any trace left in my tent or mess-basket, to remind me afterwards of the early morning visitor at my camp on Cottonwood creek."—Philadelphia, June 24th, 1851.

We have heard of some cases in which this Skunk, having penetrated into the tents of both officers and men, on our southwestern frontier, has been less skilfully managed, and the consequences were so bad as to compel the abandonment of even the tents, although soused into creeks and scrubbed with hopes of destroying the "hogo."

GEOGRAPHICAL DISTRIBUTION.

This species exists on the western ranges of the mountains in Mexico. The specimen described by LICHTENSTEIN was obtained by Mr. DEPPE in the mountains to the northwest of the city of Mexico. The animal was seen by Col. G. A. McCALL in New Mexico, between Los Vegas and Santa Fé. The specimen figured by JOHN W. AUDUBON was obtained near San Antonio, and he describes it as common in the western parts of Texas. It is not found in Louisiana, nor near the sea-shore in Texas. It will, we think, be found to inhabit some portions of California, although we cannot state this with certainty.

GENERAL REMARKS.

There are several species of this genus, which are found to vary so much in the distribution of their colours that many mere varieties were described as new species, without any other characters than those presented by the number of stripes on the back, or the predominance either of black or white spots on the different portions of the body. BUFFON described five species. Baron CUVIER, in his "Ossemens Fossiles," took much pains in endeavoring to clear up the difficulties on the subject of these animals · yet, owing to his not possessing specimens, and his too great dependence on colour, he multiplied the number of some species which are now found to be mere varieties, and omitted others which are unquestionably true species.

Hoary Marmot. The Whistler

Plate CIII

ARCTOMYS PRUINOSUS.—Pennant.

Hoary Marmot.—The Whistler.

PLATE CIII.—Males.

A. vellere cano longo, denso, maxime in thorace humorisque, in partibus posterioribus fulvo-flavescente, cauda comosa fusco nigriscente.

CHARACTERS.

Fur, long, dense, and hoary, particularly on the chest and shoulders; hinder parts dull yellowish-brown; tail bushy, blackish-brown.

SYNONYMES.

Hoary Marmot. Pennant, Hist. Quadr., vol. ii. p. 130.
   "   "   "   Arctic Zool., vol. i. p. 112.
Ground-Hog. Mackenzie's Voyage, p. 515.
Whistler. Harmon's Journal, p. 427.
Arctomys (?) Pruinosus. Rich, Zool. Jour., No. 12, p. 518. Mar. 1828.
   "   "   Rich, Fauna Boreali Americana, p. 150.
Quisquis-qui-po. Cree Indians.
Deh-ie. Cheppewyans.
Souffleur, or Mountain-Badger. Fur-Traders.
Arctomys Pruinosa. Harlan, Fauna, p. 169.
   "   Calligata. Eschscholtz, Zoologischer Atlas, Berlin, 1829, pl. 6, part 2, p. 1.

DESCRIPTION.

In form, this animal (which we examined whilst it was alive at the Zoological Gardens in London) bears a considerable resemblance to the European Marmot (*Arctomys Marmota*). It also resembles the Maryland Marmot (*A. Monax*). Being, at the time we saw it, excessively fat, the body, when it lay down, spread out or flattened like that of the badger; it was so covered with dense and very long hair that it was difficult to recognize the true outline; it subsequently shed its hair, and our figure was taken in its new and shorter pelage. The animal is rather longer than the Maryland Marmot; head, of moderate size; eyes, rather small but conspicuous; ears, oval and covered with hair on both surfaces; feet short, robust, and clothed with hair; nails strong, slightly arched, free;

tail, short, and thickly clothed with long and coarse hair to the extremity. The pelage is a soft and dense fur beneath, covered with longer and more rigid hairs.

COLOUR.

Fur on the back, dark at base, the outer portion white, with black points more or less extended; on the rump it is dull-brown at the roots, with black and yellow towards the extremities. The general appearance of the animal, owing to the admixture of these dark-brown and white hairs, of which the white predominate, is hoary-brown.

Upper surface of nose, ears, back part of the head, feet, and nails, black; a black band runs backwards from behind the ears for about an inch and a half, and then descends nearly vertically on the neck, where it vanishes; sides of muzzle, and behind the nostrils above, as well as chin, pure white; cheeks, grizzled with rust-colour and black; moustaches, nearly all black, a few, light-brown.

There are a few white hairs on the middle toes of the fore-feet; tail black, varied with rusty-brown, and a few whitish hairs with black points; whole under parts pale rust colour, with a slight mixture of black on the belly; extremities of the ears slightly tipped with white; upper incisors, yellow; lower, nearly white.

DIMENSIONS.

	Foot.	Inches.	Lines.
Length from point of nose to root of tail, - -	1	7	
" of tail (vertebræ), - - - - -		5	6
" " including hair, - - - -		7	9
Point of nose to end of head, - - - -		3	4
Ear, - - - - - - - - -			5½
Palm and nail, - - - - - - -		2	9
Nail, - - - - - - - - -			9
Tarsus, - - - - - - - - -		3	8
Nail on hind foot, - - - - - - -			8

HABITS.

This Marmot was described by PENNANT, from a skin preserved in the Leverian Museum, which was for many years the only specimen in any known collection. It appears to have afterwards become a question whether there was such an animal, or whether it might not prove to be the

Maryland Marmot, the original specimen, above mentioned, having been lost. HARLAN says of it, "This specimen was supposed to have come from the northern parts of North America." GODMAN does not mention it. Dr. RICHARDSON quotes PENNANT's description, and states that he did not himself obtain a specimen; but "if correct" in considering it as the same as the Whistler of HARMON, "we may soon hope to know more of it, for the traders who annually cross the Rocky Mountains from Hudson's Bay to the Columbia and New Caledonia are well acquainted with it." He also mentions that one, (HARMON's Whistler, we presume) which was procured for him by a gentleman, was so much injured that he did not think it fit to be sent." The Doctor then gives the following account of it, and appears to have been quite correct in supposing it identical with the animal referred to by HARMON: "The Whistler inhabits the Rocky Mountains from latitude 45° to 62°, and probably farther both ways: it is not found in the lower parts of the country. It burrows in sandy soil, generally on the sides of grassy hills, and may be frequently seen cutting hay in the autumn, but whether for the purpose of laying it up for food, or merely for lining its burrows, I did not learn. While a party of them are thus occupied, they have a sentinel on the lookout upon an eminence, who gives the alarm on the approach of an enemy, by a shrill whistle, which may be heard at a great distance. The signal of alarm is repeated from one to another as far as their habitations extend. According to Mr. HARMON, they feed on roots and herbs, produce two young at a time, and sit upon their hind-feet when they give their young suck. They do not come abroad in the winter."

"The Indians take the Whistler in traps set at the mouths of their holes, consider their flesh as delicious food, and, by sewing a number of their skins together, make good blankets."

Our drawing of this Marmot was made from the specimen now in the museum of the Zoological Society of London, which is, we believe, the only one, even at this day, to be found in Europe, with the exception of a "hunter's skin" (i. e., one without skull, teeth, or legs), which was presented to the British Museum by Dr. RICHARDSON, and was probably the one he refers to in the extract we have given above from the Fauna Boreali Americana. The specimen in the Zoological Museum is well preserved, the animal, which was alive when presented to the Society by B. KING Esq., having died in the Menagerie (Zoological Gardens) in Regent's Park.

The living animal, when we observed it, seemed to be dull and sleepy. Its cage was strewed with grass and herbs, on which it had been feeding.

GEOGRAPHICAL DISTRIBUTION.

The first specimen of this species was brought to England from Hudson's Bay. The specimen we have figured was obtained on Captain BACK's expedition. It inhabits the Rocky Mountains from 45° to 62°, and will probably be found both to the north and south of these latitudes.

GENERAL REMARKS.

It is somewhat remarkable that an animal so large as the Hoary Marmot—so widely diffused throughout the fur countries, where it is seen by traders and hunters—should be so little known to naturalists. When the living animal was brought to the Zoological Gardens it excited much interest, as the existence of the species had for many years been doubted.

We spent an hour at the Museum of the Zoological Society in London with Dr. RICHARDSON and Mr. WATERHOUSE, examining the specimen to which ESCHSCHOLZ had given the name of *A. Calligata;* and we unanimously came to the conclusion that it was the *A. Pruinosus.*

On Stone by Wm E Hitchcock

Collies Squirrel

Drawn from Nature by J.W Audubon.

Lith. Printed & Col.d by J.T. Bowen, Phil.

SCIURUS COLLIÆI.—Rich.

PLATE CIV.—Males.

S. Supra e fresco-nigro flavoque varius subtus ex flavescente albidus ; magnitudine S. migratorii.

CHARACTERS.

Size of Sciurus Migratorius ; upper parts mottled brownish black and yellow ; under surface cream white.

SYNONYMES.

Sciurus Colliæi. Richardson, Append. to Beechey's Voyage.
"   "   Bachman, Proc. Zool. Soc. 1838 (Monog. of Genus Sciurus).

DESCRIPTION.

In size and form this species bears some resemblance to the migratory gray Squirrel of the middle or northern States ; the tail, however, in the only specimen which exists in any collection, appears much smaller and less distichous, and the animal, when other specimens are examined, may prove to be intermediate in size between the Carolina gray Squirrel and *S. Migratorius.*

The fur is rather coarse, and the tail appears to be somewhat cylindrical ; ears, of moderate size, ovate, clothed with short hairs on both surfaces, but not tufted.

COLOUR.

Above, grizzled with black and dull yellow ; sides of the muzzle, under parts of the body, and inner sides of limbs, dull-white ; tail, moderate, the hairs grayish-white, three times annulated with black. Hairs of the body, both above and beneath, grey at the roots, those on the back having lengthened black tips broadly annulated with dull-yellow. The hairs of the head resemble those of the back, except on the front, where they are annulated with dull-white ; top of the muzzle, brown ; cheeks, greyish ;

insides of ears, yellowish, indistinctly freckled with brown; outsides, grizzled with black and yellow on the forepart, but posteriorly covered with long whitish hairs; hairs on the feet, black at the roots, white at the tips, the feet and legs being dirty cream-colour, pencilled with dusky; whiskers, long as the head, composed of bristly black hairs. The above description was taken by us from the specimen in the Zoological Society's Museum, London; the skin was not in very good condition, and a portion of the tail was wanting.

DIMENSIONS.

	Inches.	Lines.
Length from nose to root of tail,	10	9
" of tail to end of hair,	9	6
Height of ear posteriorly,		6
Tarsus (including nail),	2	5
Nose to ear,	2	0

HABITS.

Our figures of this Squirrel were made from the specimen presented to the Zoological Society of London by Captain BEECHEY; the original from which the species was described and named by our friend Dr. RICHARDSON.

All the information we have as to the habits of this animal is contained in the above-mentioned appendix (p. 8): "Mr. COLLIE observed this Squirrel, in considerable numbers, sporting on trees at San Blas in California (?), where its vernacular name signifies 'Little Fox-Squirrel.' It feeds on fruits of various kinds. Although unwilling to incur the risk of adding to the number of synonymes with which the history of this large genus is already overburdened, I do not feel justified in referring it to any of the species admitted into recent systematic works; and I have therefore described it as new, naming it in compliment to the able and indefatigable naturalist who procured the specimen."

GEOGRAPHICAL DISTRIBUTION.

This species was given by RICHARDSON, as appears by the above quotation, as existing at San Blas, California; this place, however, if we have not mistaken the locality, is in the district of Xalisco in Mexico, and within the tropics; it is doubtful, therefore, whether the species will be found to inhabit any portion of California. J. W. AUDUBON did not observe it in his travels through Upper California.

GENERAL REMARKS.

This species is very nearly allied to *Sciurus Aureogaster* of F. CUVIER, and it is yet possible that it may prove a variety of that very variable species, in which the under parts of the body are sometimes white, instead of the usual deep-red colour.

A specimen of *S. Aureogaster* in the Museum at Paris has the under parts of the body white, with small patches of red, and with a few scattered red hairs here and there mingled with the white ones.

PSEUDOSTOMA DOUGLASII.—Rich.

Columbia Pouched-Rat.

PLATE CV.—Males.

P. Supra fusca, lateribus subrufis, ventre pedibusque pallidioribus, cauda corporis dimidio longiore.

CHARACTERS.

Above, dusky brown ; reddish on the sides ; paler beneath and on the feet ; tail exceeding half the length of the body.

SYNONYME.

Geomys Douglasii. Richardson, Columbia Sand-Rat, Fauna Boreali Americana, p. 200, pl. 18 B.

DESCRIPTION.

Head, large and depressed ; ears, short, ovate, extending beyond the fur ; nose, blunt ; nostrils, small and round, separated by a line in the septum ; they have a small naked margin. Mouth, of moderate size ; lips, and space between the nose and upper incisors, covered with short hair ; incisors strong, and slightly recurved ; upper ones with a distinct furrow on the anterior surface, near their inner edge ; cheek pouches, large, opening externally (like those of all the other species belonging to this genus), and lined on the inside with very short hairs.

The pouches extend from beneath the lower jaw along the neck to near the shoulders ; whiskers, short ; body cylindrical, resembling that of the mole, and covered with short, dense, velvety fur ; the tail, which is round and tapering, although at first sight appearing naked, is covered with hair throughout its whole length, but most densely near the root ; legs short, and moderately robust ; fore-toes short, the three middle ones united at their base by a skin, the outer one smaller and farther back ; thumb, very small and armed with a claw ; claws, sharp-pointed, compressed, and slightly curved ; palms naked, and on the posterior part filled by a large, rounded callosity. The palms in this species are much smaller than in *P. Bursarius* ; the hind-feet are rather more slender than

Plate CV.

Drawn from Nature by J.W. Audubon.

On Stone by W.E. Hitchcock

Lith. Printed & Col.d by J.T. Bowen, Philad.a

Columbia Pouched Rat.

the fore-feet, and their claws are decidedly smaller; soles of hind-feet, entirely naked, and without any conspicuous tubercles; heel, naked, and narrow; feet and toes, thickly clothed with hair extending to the nails.

COLOUR.

Incisors, dull orange; whiskers, nearly all white; upper surface of body, top of the head, and along the sides of the pouches, dusky-brown; sides, reddish-brown; edges of pouches, dark-brown; under surface of body, feet, and tail, pale buff; nails, yellowish-white.

DIMENSIONS.

	Inches.	Lines
Length of head and body. - - - - - -	6	6
" head, - - - - - - -	1	10
" tail (vertebræ), - - - - - -	2	10
From point of nose to eye, - - - - -		11
" " " auditory opening, - - -	1	8
Between the eyes, - - - - - - -		7
From wrist joint to end of middle claw, - - -	1	

HABITS.

This species of Sand-Rat was first obtained by Mr. DAVID DOUGLAS. near the mouth of the Columbia river, since which, specimens have been sent to England by various collectors. According to Mr. DOUGLAS, the animal, "when in the act of emptying its pouches, sits on its hams like a Marmot or Squirrel, and squeezes its sacs against the breast with the chin and fore-paws."

"These little Sand-Rats are numerous in the neighbourhood of Fort Vancouver, where they inhabit the declivities of low hills, and burrow in the sandy soil. They feed on acorns, nuts (*Corylus rostrata*), and grasses, and commit great havoc in the potato-fields adjoining to the fort, not only by eating the potatoes on the spot, but by carrying off large quantities of them in their pouches."—*Fauna Boreali Americana*, p. 201.

GEOGRAPHICAL DISTRIBUTION.

This species inhabits the valleys to the west of the Rocky Mountains, and seems to have been most frequently observed in about the latitude of the mouth of the Columbia River. Its probable range may extend as far as California to the south, and the Russian Possessions in the opposite direction. We have seen some mutilated specimens, which appeared to be of this species, obtained by a party in the western portion of New Mexico, but so dilapidated were they, that it was impossible to decide positively as to their identity, and they may have been skins of another species, called by Dr. RICHARDSON *Geomys Umbrinus,* which he was informed came from the southwestern part of Louisiana.

GENERAL REMARKS.

Mr. DOUGLAS informed Dr. RICHARDSON "that the outside of the pouches was cold to the touch, even when the animal was alive, and that on the inside they were lined with small, orbicular, indurated glands, more numerous near the opening into the mouth. When full, the pouches had an oblong form, and when empty they were corrugated or retracted to one third of their length."

We presume this information is correct, although the mistake made by supposing the "inverted" pouches of some species of Pseudostoma, to be in their natural position (see the genus *diplostoma* of RAFFINESQUE, adopted by RICHARDSON), leads us to look with caution on any accounts of the pouches of our Sand-Rats from this source.

Plate CVL.

Drawn from Nature by J.W. Audubon

On Stone by W. E. Hitchcock

Columbian Black Tailed Deer.

Lith. Printed & Col.ᵈ by J.T. Bowen, Philadª

CERVUS RICHARDSONII.—Aud. and Bach.

Columbian Black-tailed Deer.

PLATE CVI.—Males.

C. Supra subrufus, infra albus, auriculis mediocribus, angustioribus quam in C. macrotide, corpore minore, ungulis angustioribus et acutioribus quam in uto, macula albida in natibus nulla, cornibus teretibus bis bifurcatis.

CHARACTERS.

Ears, moderate, narrower than in C. Macrotis ; *size, less than* C. Macrotis ; *hoofs, narrower and sharper ; no light patch on the buttocks ; colour, reddish-brown above, white beneath ; horns, cylindrical, twice bifurcated.*

SYNONYMES.

Cervus Macrotis. Rich (non Say) Black-tailed Deer, Fauna Boreali Americana, p. 254, pl. 20.
California Deer, of gold diggers.

DESCRIPTION.

Male.—In size this animal a little exceeds the Virginian Deer, but it is less than the Mule Deer (*C. Macrotis*) ; in form it is shorter and stouter than *C. Virginianus.*

There is a tuft of long pendulous hairs hanging down from the umbilicus backward to between the thighs. The horns are nearly cylindrical, and are twice forked ; the first bifurcation being ten inches from the base—about five to six inches longer to that fork than in *C. Macrotis,* as described by Say. There is a knob, in the specimen from which we describe, on one horn, about four inches from the base ; the horn continues in a single branch for about ten inches, where it divides into two branches, each of which has two points ; and the antlers may be said to bear some resemblance to those of the Red Deer of Europe, much greater than do those of the Virginian Deer or Elk.

Ears, of moderate size ; head, proportionately a little shorter than the head of the Virginian Deer and nose less pointed ; hoofs, narrow and sharp, and longer and more pointed than those of the **Mule Deer** (*C. Macrotis*), which are round and flattened.

The lachrymal openings are large, and situated close beneath the eye ; tail, rather short, stouter and more bushy than that of *C. Macrotis.*

COLOUR.

A brown mark originating between the nostrils is continued behind their naked margins, downwards, towards the lower jaw, uniting with a dark patch situated behind the chin ; chin and throat, white ; forehead, dark-brown ; neck, back, sides, and hips, brownish-gray ; hairs clothing those parts, brown from their roots to near their tips, where they exhibit a pale yellowish-brown ring surmounted by a black tip ; on the back part of the neck there is a dark line down the middle of the back, becoming lighter as it recedes from the neck.

The chest is blackish-brown, running around the shoulder somewhat like the mark of a collar ; a dark line extends from under the chest to the centre of the belly ; the anterior of the belly is fawn-coloured, the posterior part white, as are likewise the insides of the thighs ; the tail, at its junction with the back, is dark brown, and this colour increases in depth to the tip, which is black ; the under side of the tail is clothed with long white hairs ; the legs are mixed yellowish-brown and black anteriorly, and pale brownish-white posteriorly.

DIMENSIONS.

	Feet.	Inches
Length from tip of nose to brow (between the horns),	1	
" " " to root of tail, - - -	5	4
" of tail (vertebræ), - - - - - -		6
" " (to end of hair), - - - - -		9
Height at shoulder, - - - - - - -	2	6
Width of horns between superior prongs, - - -	1	8
" " " posterior pair of points, - -	1	3

HABITS.

This beautiful Deer is found variously dispersed over the western portions of the North American continent, where it was first noticed by LEWIS and CLARK, near the mouth of the Columbia River ; but not until the discovery of the golden treasures of California did it become generally known to white men. In that country, along the hill sides and in the woody dells and "gulches," the hardy miners have killed hundreds, nay thousands, of Black-tailed Deer ; and it is from the accounts they have given that it is now known to replace, near the great Sierra Nevada, the common or Virginian Deer which is found east of the Rocky Mountains ; all the hunters who have visited California, and whom we have seen, tell us that every Deer they shot there was the Black-tailed species.

J. W. AUDUBON killed a good many of these Deer, and describes them as tender and of good flavour ; and during the time his party encamped on the Tuolome River, and in the "dry diggings" near Stockton, when he kept two of his men busy shooting for the support of the others, they generally had one or two Deer brought into camp every day. The mode of hunting them was more similar to what is called Deer-stalking in Scotland than to the methods used for killing Deer in the eastern part of the Union. Sometimes the hunters (who had no dogs) would start before day, and, gaining the hills, anxiously search for fresh tracks in the muddy soil (for it was then the rainy season, and the ground everywhere wet and soft), and, having found a trail, cautiously follow ; always trying to keep the wind in such a direction as not to carry the scent to the animals. After discovering a fresh track, a search of a most tedious and toilsome nature awaited them, as the unsuspecting Deer might be very near, or miles off, they knew not which ; at every hill-top they approached, they were obliged to lie down and crawl on the earth, pausing when they could command the view to the bottom of the valley which lay beyond the one they had just quitted ; and after assuring themselves none were in sight, carefully following the zigzag trail, proceed to the bottom. Again another summit has been almost reached ; now the hunters hope for a shot : eye and ear are strained to the utmost, and they move slowly forward : the ridge of the next hill breaks first upon their sight beyond a wide valley. The slope nearest them is still hidden from their view. On one side the mountains rise in steeper and more irregular shapes ; pine-trees and oaks are thickly grown in the deepest and most grassy spot far below them. The track trends that way, and silently they proceed, looking around at almost every step, and yet uncertain where their game has wandered. Once the trail has been almost lost in the stony, broken ground they pass, but again they have it ; now they approach and search in different directions the most likely places to find the Deer, but in vain ; at last they gain the next summit : the object of their chase is at hand ; suddenly they see him—a fine buck—he is yet on the declivity of the hill, and they cautiously observe his motions. Now they see some broken ground and rocky fragments scattering towards the left ; they redouble their caution ; locks are ready cocked ; and, breathing rapidly, they gain the desired spot

One instant—the deadly rifle has sent its leaden messenger and the buck lies struggling in his gore.

Short work is made of the return to camp if no more Deer signs are about; and a straight cut may bring the hunters home in less than an hour, even should they have been two or three in following their prize.

Sometimes the Deer start up suddenly, quite near, and are shot down on the instant; occasionally, after a long pursuit, the crack of a rifle from an unknown hunter deprives the others of their chance; and—must we admit it—sometimes they miss; and not unfrequently they see no game at all.

Mr. J. G. BELL informed us that while he was digging gold in a sequestered and wild cañon, in company with a young man with whom he was associated in the business, they used to lie down to rest during the heat of the day, and occasionally he shot a Black-tailed Deer, which unsuspectingly came within shooting distance down the little brook that flowed in the bottom of the ravine. He also used to rise very early in the mornings occasionally, and seek for the animals in the manner of still-hunting, as practised in the United States. One morning he killed three in this manner, before his breakfast-time, and sold them, after reserving some of the best parts for himself and companion, for eighty dollars apiece! He frequently sold Deer subsequently, as well as hares and squirrels, birds, &c., which he shot at different times, for enormous prices. Many of the miners, indeed, turned their attention to killing Deer, elk, bear, antelopes, geese, ducks, and all sorts of game and wild fowl, by which they realized considerable sums from selling them at San Francisco and other places. We have heard of one person who, after a luckless search for gold, went to killing Deer and other game, and in the course of about eighteen months had made five thousand dollars by selling to the miners at the diggings.

The gait of this species is not so graceful as that of the Virginian Deer; it bounds rather more like the roebuck of Europe than any other of our Deer except the Long-tailed Deer, and is reported to be very swift. The season of its breeding is earlier than that of the common Deer, and it no doubt brings forth the same number of young at a time.

GEOGRAPHICAL DISTRIBUTION.

This beautiful Deer was first met with by J. W. AUDUBON on the eastern spurs of the coast range of mountains after leaving Los Angeles and traversing a portion of the Tule valley in California. On entering the broad plain of the San Joaquin and river of the lakes, few Black-tailed Deer were met with, and the elk and antelope took their place. The party again found them abundant when they reached the hills near the Sierra Nevada, on their way towards the Chinese diggings, about eighty miles southeast of Stockton.

They may be said to inhabit most of the hilly and undulating lands of California, and as far as we can judge probably extend on the western side of the grand ridge of the Rocky Mountains nearly to the Russian Possessions.

We have not heard that they are met with east of the bases of that portion of the Cordilleras which lies in the parallel of San Francisco, or north or south of that latitude, although they may exist in the valleys of the Colorado of the west in a northeast direction from the mouth of that river, which have as yet not been much explored.

GENERAL REMARKS.

According to our present information, there is only one specimen of this Deer in the collections of objects of natural history in Europe, and this is in the museum of the Zoological Society in London, where it was, when we saw it, (erroneously) labelled *C. Macrotis.*

At the Patent Office in Washington city there is a skin of a Deer (one of the specimens brought from the northwest coast of America by the Exploring Expedition), which has been named by Mr. PEALE *C. Lewisii.*

We have not positively ascertained whether it be distinct from our *C. Richardsonii,* but presume it will prove to be well separated from it, as well as from all our hitherto described Deer, and we shall endeavour to figure it, if a good species, and introduce it into our fauna under the name given it by Mr. PEALE.

We have detected an error in the description of the horns of *C. Macrotis* (see vol. ii. p. 206), where a portion of the description of those of *C. Richardsonii* seems to have been introduced by mistake.

ARCTOMYS LEWISII.—Aud. and Bach.

LEWIS'S MARMOT.

PLATE CVII.—Males.

A. Rufo-fulvus, pedibus albo-virgatis, cauda apice albo; magnitudine leporis sylvatici, forma a monacis.

CHARACTERS.

Size of the grey rabbit; general shape of the head and body similar to that of A. monax; *colour reddish-brown; feet barred with white; end of tail white.*

DESCRIPTION.

Head, rather small; body, round and full; ears short, ovate, with somewhat acute points, thickly clothed with short hairs on both sides: whiskers long, extending beyond the ears; nose blunt, naked; eyes, of moderate size; teeth, rather smaller than those of the Maryland marmot; feet, short; nails, rather long and arched, the nail on the thumb being large and nearly the size of the others; tail short, round, not distichous, thickly clothed with hair to the end; the hair is of two kinds—a short, dense fur beneath, with longer and rigid hairs interspersed.

COLOUR.

Nose, black; incisors, yellowish-white; nails, black; the whole upper surface and the ears, reddish-brown; this colour is produced by the softer fur underneath being light yellowish-brown, and the longer hairs, at their extremities, blackish-brown. On the haunches the hairs are interspersed with black and yellowish-brown; feet and belly, light salmon-red; tail, from the root for half its length, reddish-brown, the other half to the tip soiled white; above the nose, edges of ears, and along the cheeks, pale reddish-buff.

There is a white band across the toes, and another irregular one behind them; and an irregularly defined dark-brown line around the back of the head and lower part of the chin, marking the separation of the head from the throat and neck.

Plate CVII

Drawn from Nature by J. W. Audubon.

Drawn on Stone by Wm E. Hitchcock

Lith Printed & Col. by J.T.Bowen Philᵃ.

Lewis Marmot.

DIMENSIONS.

	Feet.	Inches
From nose to root of tail, - - - - - -	1	4
Tail (vertebræ), - - - - - - - -		2
" (to end of hair), - - - - - - -		8
Point of nose to ear, - - - - - - -		2
" " to eye, - - - - - - -		1
Heel to middle claw, - - - - - - -		2½

HABITS.

From the form of this animal we may readily be convinced that it possesses the characteristics of the true Marmots. These animals are destitute of cheek-pouches ; they burrow in the earth ; live on grasses and grains ; seldom climb trees, and when driven to them by a dog do not mount high, but cling to the bark, and descend as soon as the danger is over. As far as we have been able to ascertain, all the spermophiles or burrowing squirrels are gregarious, and live in communities usually numbering several hundreds, and often thousands. On the contrary, the Marmots, although the young remain with the mother until autumn, are found to live solitarily, or at most in single pairs. It was not our good fortune ever to have met with this species in a living state, hence we regret that we are unable to offer anything in regard to its peculiar habits.

GEOGRAPHICAL DISTRIBUTION.

We have no doubt this species, like the other Marmots, has an extensive geographical range, but coming from so distant a part of our country as Oregon, which has been so little explored by naturalists, we are obliged to make use of the vague term "shores of the Columbia river" as its habitat.

GENERAL REMARKS.

We have not felt at liberty to quote any authorities or add any synonymes for this species, inasmuch as we cannot find that any author has referred to it. The specimen from which our figure was made, and which we believe is the only one existing in any collection, was sent to the Zoological Society by the British fur-traders who are in the habit of annually carrying their peltry down the Columbia river to the Pacific. It is labelled in the museum of the Zoological Society, No. 461, page 48 Catalogue, *Arctomys brachyura ?* HARLAN. The history of the supposed species of HARLAN is the following : LEWIS and CLARK (Expedition, vol. ii. p. 173) describe an animal from the plains of the Columbia under the name of burrowing squirrel. No specimen was brought. HARLAN and RAFINESQUE in quick succession applied their several names to the species, the former styling it *Arctomys brachyura* and the latter *Anisonyx brachyura*. When the present specimen was received at the Museum, the name of *A. brachyura* was given to it, with a doubt. On turning to LEWIS and CLARK's descriptions, the only guides which any naturalists possess in reference to the species, we find that they refer to an animal whose whole contour resembles that of the squirrel, the thumbs being remarkably short and equipped with blunt nails, and the hair of the tail thickly inserted on the sides only, which gives it a flat appearance, whereas the animal of this article does not resemble a squirrel in its whole contour ; its thumbs, instead of being remarkably short and equipped with blunt nails, have long nails nearly the length of those on the other toes, and the tail, instead of being flat with the hairs inserted on the sides, is quite round. It differs also so widely in several other particulars that we deem it unnecessary to institute a more minute comparison. We have little doubt that LEWIS and CLARK, who, although not scientific naturalists, had a remarkably correct knowledge of animals, and described them with great accuracy, had, in their account of the burrowing squirrel, reference to some species of spermophile—probably *Spermophilus Townsendii*, described in this volume— which certainly answers the description referred to much nearer than the species of this article.

Plate CVIII

Drawn from Nature by J.W. Audubon.

On Stone by W.E. Hitchcock

Bachman's Hare.

Lith Printed & Col.d by J.T. Bowen Philad.a

LEPUS BACHMANI.—Waterhouse.

BACHMAN'S HARE.

PLATE CVIII.—Males.

L. Supra fuscus, lateribus cinereo fuscis, ventre albo rufo-tincto ; L. sylvatico aliquantulo minor, auriculis capite paullo longioribus.

CHARACTERS.

A little smaller than the gray rabbit ; ears rather longer than the head ; tarsi, short. Colour, brown above, gray-brown on the sides, belly white, tinged with rufous.

SYNONYMES.

LEPUS BACHMANI. Waterhouse, Proceedings Zool. Soc. 1838, p. 103.
" " Bachman's Hare, Bach. Jour. Acad. Nat. Sci. Phila., vol. viii. part 1, p. 96.
" " Waterhouse, Nat. Hist. Mamm., vol. ii. p. 124.

DESCRIPTION.

This Hare bears a general resemblance to the gray rabbit (*L. sylvaticus*), but is considerably smaller : the fur is softer and the ears shorter than in that species.

Upper incisors, much arched, and deeply grooved ; claws, slender and pointed—the claw of the longest toe remarkably slender ; ears longer than the head, sparingly furnished with hair quite fine and closely adpressed externally ; tail, short ; feet, thickly clothed with hair covering the nails.

COLOUR.

The fur on the back and sides is deep gray at the roots, annulated near the ends of the hairs with brownish-white, and black at the points. On the belly the hair is gray at the roots and white at the points, with a tinge of red ; chest and fore parts of the neck, gray-brown, each hair being dusky at the tip ; chin and throat, grayish-white ; the hairs on the head are brownish-rufous; on the flanks there is an indistinct pale longitudinal dash just above the haunches; under surface of tail white, edged with brownish-black; general colour of the tarsus above, dull-rufous; sides of tarsus, brown; ears, on the fore part mottled with black and yellowish-white, on the hinder part greyish-white; internally the ears are dull orange, with a white margin all around their openings; their apical portion is obscurely margined with black.

DIMENSIONS.

	Inches.	Lines.
Length from point of nose to root of tail, - - -	10	
Tail (vertebræ),- - - - - - -		9
" to end of fur, - - - - - - -	1	3
Ear internally, - - - - - - - -	2	8
From heel to point of longest nail, - - - -	3	
Tip of nose to ear, - - - - - - -	2	5

HABITS.

The manners of this pretty Hare, as observed in Texas by J. W. AUDUBON, appear to assimilate to those of the common rabbit (*Lepus sylvaticus*), the animal seldom quitting a particular locality, and making its form in thick briar patches or tufts of rank grass, keeping near the edges of the woody places, and being seen in the evenings, especially for a short time after sunset, when it can be easily shot.

We have been favoured with the following particulars as to the habits of this Hare by our esteemed friend Captain J. P. McCOWN of the United States Army:

"This Hare is deficient in speed, and depends for its safety upon dodging among the thick and thorny chaparals or nopal clusters (*cacti*) which it inhabits, never venturing far from these coverts.

"Large numbers can be seen early in the morning or late in the evening, playing in the small openings or on the edges of the chaparals, or nibbling the tender leaves of the nopal, which seems to be the common prickly pear of our country, only much larger from congeniality of climate."

"The principal enemies of these Hares in Texas are the cat species, hawks, and snakes."

During the war with Mexico, some of the soldiers of our army who were stationed on the Mexican frontier had now and then a sort of battue, to kill all the game they could in their immediate vicinity; and by surrounding a space of tolerably open ground, especially if well covered with high grass or weeds, and approaching gradually to the centre, numbers of these Hares were knocked down with clubs as they attempted to make their escape, as well as occasionally other animals which happened to be secreted within the circle. We were told that a raw German recruit, who had once or twice before been made the butt of his comrades, having joined only a few days, was invited to partake of the sport, and as the excitement became quite agreeable to him, was amongst the foremost in knocking down the unfortunate Hares, as they dashed out or timidly squatted yet a moment, hoping not to be observed ; when suddenly one of his companions pointed out to him a *skunk*, which, notwithstanding the din and uproar on all sides, was very quietly awaiting the course of events. The unlucky recruit darted forward :—we need say nothing more, except that during the remainder of the war the skunk was, by that detachment, known only as the "Dutchman's rabbit."

This Hare so much resembles the common rabbit, that it has been generally considered the same animal ; and this is not singular, for the gray rabbit does not extend to those portions of our country in which BACHMAN'S Hare is found, and few, save persons of some observation, would perceive the differences between them, even if they had both species together so that they could compare them.

GEOGRAPHICAL DISTRIBUTION.

Lieut. ABERT, of the United States Army, procured specimens of this Hare in the neighbourhood of Santa Fé, which were the first that were made known to naturalists as existing east of California, as the animal was described from a specimen sent by DOUGLAS from the western shores of America. It now appears that it occupies a great portion of Texas, New Mexico, and California, probably extending south through great part of Mexico. Its northeastern limit may be about the head waters of the Red river or the Arkansas.

GENERAL REMARKS.

From the small size of this Hare, it was at one time considered possible that it might prove to be only the young of some other species of *Lepus*, but its specific characters are now fully established, and it is, at present, known as more numerous in some localities than even the gray rabbit.

This species was discovered among a collection of skins in the museum of the Zoological Society by Dr. BACHMAN and Mr. WATERHOUSE, and the latter gentleman having desired the doctor to allow him to describe and name it, called it *L. Bachmani*, in compliment to him. Our figures were made from the specimen described by Mr. WATERHOUSE, which is yet in the museum of the Zoological Society at London. We have obtained many skins since, from Texas and the southwestern portions of New Mexico.

Plate CIX

N 22

Drawn from Nature by J.W.Audubon.

Drawn on Stone by Wm E Hitchcock

Lith. Printed & Col.d by J.T. Bowen, Phil.a

Mexican Marmot-Squirrel

Adult male and young

SPERMOPHILUS MEXICANUS.—Licht.

MEXICAN MARMOT SQUIRREL (SPERMOPHILE).

PLATE CIX.—OLD MALE, and YOUNG.

S. magnitudine sciuri Hudsonici, auriculis brevibus, cauda longa, corpore supra rufo-fulvo, maculis vel strigio albis, subtus albo flavescente.

CHARACTERS.

Size of Sciurus Hudsonicus ; *ears, short ; tail, long ; body, above, reddish-tawny, with white spots or bars ; beneath, yellowish-white.*

SYNONYMES.

CITILUS MEXICANUS. Licht., Darstellung neuer oder wenig bekannter Säugthiere, Berlin, 1827–1834.
SPERMOPHILUS SPILOSOMA. Bennett, Proc. Zool. Soc., London, 1833, p. 40.

DESCRIPTION.

Form, very similar to the leopard spermophile (*S. tridecemlineatus*), although the present species is the larger of the two ; ears, short, and clothed with short hairs ; body, moderately thick ; legs, rather short ; toes and nails, long ; tail, somewhat flat, distichous. and shorter than the body.

COLOUR.

Upper surface, rufous-brown, spotted with yellowish-white, the spots bordered posteriorly with black ; under parts, pale buff-white ; this colour extends somewhat upwards on the sides of the animal ; feet, pale-yellow ; tarsi, hairy beneath, the hairs extending forwards to the naked fleshy pads at the base of the toes ; claws, dusky horn colour, with pale points ; the fur at the roots (both on the upper and under parts of the animal) is gray.

The eye is bordered with whitish-yellow ; head and ears, rufous-brown ; upper surface of tail, dark-brown, edged with a white fringe on the sides ;

towards the extremities the hairs are yellow, but they have a broad black band in the middle of their length ; under surface of the tail of an almost uniform yellowish-hue, slightly inclining to rust colour.

DIMENSIONS.

Adult male.

	Inches.	Lines.
From point of nose to root of tail, - - -	10	
Tail (vertebræ), - - - - - - -	4	
" including hair, - - - - - - -	5	
Nose to end of head, - - - - - -	2	6
Length of ears, - - - - - - -		4
From elbow of fore-leg to end of longest nail, -	2	6
Tarsus (of hind leg), - - - - - -	1	9

Measurements of the specimen named *S. Spilosoma* by Mr. BENNETT :

Young.

	Inches.	Lines
From point of nose to root of tail, - - -	5	9
Tail (vertebræ), - - - - - -	2	9
" including hair, - - - - - -	3	6
Nose to ear, - - - - - - -	1	3
Tarsus and nails, - - - - - -	1	3
Length of nail of middle toe, - - - -		2½
" fore foot and nails, - - - -		9½
" middle toe of fore foot to nail, - -		2½

HABITS.

This Mexican Spermophile has all the activity and sprightliness of the squirrel family, and in its movements greatly reminds one of the little ground-squirrel (*Tamias Lysteri*) of the middle and northern States. It feeds standing on its hind feet and holding its food in the fore paws like a common squirrel, and is remarkable for the flexibility of its back and neck, which it twists sideways with a cunning expression of face while observing the looker on. When caught alive this pretty species makes a pet of no common attractions, having beautiful eyes and being very handsomely marked, while its disposition soon becomes affectionate, and it retains its gay and frolicsome habits. It will eat corn and various kinds of seeds, and is fond of bits of potatoe, apple, or any kind of fruit, as well as bread, pastry, cakes, &c.: grasses and clover it will also eat readily, and in fact it takes any kind of vegetable food. Even in the hottest summer weather this animal is fond of making a nest of tow and bits of carpet, and will sleep covered up by these warm materials as comfortably as if the temperature was at freezing point outside instead of 85°.

For some time we have had a fine living animal of this species in a cage, and he has been a source of great amusement to the little folks, who are fond of feeding him and pleased to see his antics. When threatened he shows fight, and approaches the bars of his cage gritting or chattering with his teeth like a little fury, and sometimes uttering a sharp squeak of defiance; but when offered any good thing to eat he at once resumes his usual playful manner, and will take it from the hand of any one. In eating corn this little animal picks out the soft part and leaves the shell and more compact portion of the grain untouched.

At times he will coil himself up, lying on one side, almost entirely concealed by the tow and shredded carpet: if then disturbed, he looks up out of one eye without changing his position, and will sometimes almost bear to be poked with a stick before moving. Like the human race he occasionally shows symptoms of laziness or fatigue, by yawning and stretching. When first placed in his cage he manifested some desire to get out, and attempted to gnaw the wires: he would now and then turn himself upside down, and with his fore paws holding on to the wires above his head bite vigorously at the horizontal wires for half a minute at a time, before changing this apparently uncomfortable position. This Spermophile is not in the habit of eating a very great deal at a time, but seems to prefer feeding at intervals, even when plenty of food lies within his reach, retiring to his snug nest and sleeping for a while after eating a sufficient portion. When thus sleeping we sometimes found him lying on his back, with his fore paws almost joined, held close by his nose, while his hind legs were slightly turned to one side so as to give his body the appearance of complete relaxation.

These animals are said to be tolerably abundant in Mexico and California, but only in the wooded districts. We were informed that they could easily be procured near Vera Cruz, Tuspan, Tampico, &c.

GEOGRAPHICAL DISTRIBUTION.

LICHTENSTEIN informs us that Mr. DEPPE procured this animal in 1826, in the neighbourhood of Toluca in Mexico, where it was called by the inhabitants by the general term *Urion*, which was also applied to other burrowing animals. Captain BEECHY states that his specimen was procured in California, and we are informed by Captain J. P. McCOWN that it exists along the Rio Grande and in other parts of Texas, where he has seen it as a pet in the Mexican ranchos.

GENERAL REMARKS.

In our first edition (folio plates), we gave figures of the *young* of this species as *S. spilosoma* of BENNETT, but having since ascertained that his specimen was only the young of *S. Mexicanus*, a species which had been previously published, we have now set down *S. spilosoma* as a synonyme of the latter, and have placed the figures of both old and young on the same plate.

Plate CX

Drawn from Nature by J.W. Audubon.

On Stone by W.E.Hitchcock.

Lith. Printed & Col.ᵈ by J.T. Bowen, Philad.ᵃ

Mole Shaped Pouched Rat.

PSEUDOSTOMA TALPOIDES.—Rich.

MOLE-SHAPED POUCHED RAT.

PLATE CX.—Males.

P. **Magnitudine muris** ratti, corpore nigro cinerescente, capite pro portione **parvo,** mento albo, macula alba ad gulam, pedibus posticis **quadridigitatis.**

CHARACTERS.

Size of the black rat ; head, small in proportion ; body, grayish-black ; chin, white ; a white patch on the throat ; only four perfect toes on the hind feet.

SYNONYMES.

CRICETUS (?) TALPOIDES. Rich, Zool. Jour. No. 12, p. 5, pl. 18.
? GEOMYS ? TALPOIDES. " F. B. A., p. 204.
OOTAW-CHEE-GOES-HEES. Cree Indians.

DESCRIPTION.

Body, shaped like that of the mole : head, rather small ; nose, obtuse and covered with short hairs ; incisors, strong, with flat anterior surfaces ; upper ones short and straight, and each marked with a single very fine groove close to their inner edge ; lower incisors, long, curved inwards, and not grooved ; whiskers, composed of fine hairs as long as the head ; eyes, small ; auditory opening, small and slightly margined ; ears, scarcely visible beyond the fur.

The pouches have an opening on the sides of the mouth externally, and are of moderate size ; extremities, very short ; the fore foot has four toes and the rudiment of a thumb ; the middle toe is longest and has the largest claw, the first and third are equal to each other in length, the outer one is shorter and placed far back, and the thumb, which is still farther back, consists merely of a short claw ; the fore claws are long, compressed, slightly curved, and pointed ; they are, however, less robust than those of some other species of the genus, especially *P. bursarius.* On the hind feet there are four short toes, armed with compressed claws much shorter than those on the fore feet, and the rudiment of a fifth toe, so small that it can be detected only after a minute inspection ; tail, very slender, cylindrical, and rather short, covered with a smooth coat of short hairs.

The hair is nearly as fine as that of the common shrew mole, and is close and velvety.

COLOUR.

Whiskers, black ; incisors, yellowish-white, approaching flesh colour ; chin and throat, white ; outer edges of the pouch, light gray ; tail, grayish-brown ; the body generally, grayish-black, with faint brownish tints in some lights.

DIMENSIONS.

	Inches.	Lines.
Length of head and body, - - - - - -	7	4
Tail to end of hair, - - - - - -	2	5
From point of nose to eye, - - - - -		9
From point of nose to auditory opening, - -	1	3
Height of back, - - - - - - -	2	
Length of lower incisors, - - - - -		5
" fur on the back, - - - - -		6
" middle fore claw, - - - - -		4
From heel to end of middle hind claw, - - -		11

HABITS.

Very little is known of the habits of this peculiar sand-rat. The manners, however, of all the species of the genus *Pseudostoma* are probably very similar : they live principally under ground, and leave their galleries, holes, or burrows, pretty much as we of the genus *Homo* quit our houses, for the purpose of procuring the necessaries of life, or for pleasure, although they do find a portion of their food while making the excavations which serve them as places in which to shelter themselves and bring forth their young. They are generally nocturnal, and in the day time prefer coming abroad during cloudy weather.

They never make their appearance, nor do they work in their galleries or burrows during the winter in our northern latitudes, unless it be far beneath the hard frozen ground, which would not permit them to make new roads.

RICHARDSON says that as soon as the snow disappears in the spring, and whilst the ground is as yet only partially thawed, little heaps of earth newly thrown up attest the activity of this animal.

The specimen from which our figures were made was presented to the Zoological Society by Mr. LEADBEATER, who obtained it from Hudson's Bay. It also served Dr. RICHARDSON for his description : he was inclined to identify it with a small animal inhabiting the banks of the Saskatchewan, which throws up little mounds in the form of mole hills, but generally rather larger ; he, however, could not procure any specimens.

As an evidence that this animal never feeds upon worms, he mentions the fact that none exist in high northern latitudes. A gentleman who had for forty years superintended the cultivation of considerable pieces of ground on the banks of the Saskatchewan, informed him that during the whole of that period he never saw an earthworm turned up. All the species of *Pseudostoma*, as far as our knowledge goes, feed on bulbs, roots, and grasses.

The pouches serve as sacks, in which after filling them with food they carry it to their nests in their subterranean retreats, where they deposit considerable quantities, which evidently serve them as supplies throughout the winter.

We are under the impression that none of the species of this genus become perfectly dormant in winter, as we have observed in Georgia a few fresh hillocks thrown up by the Southern pouched-rat after each warm day in that season.

GEOGRAPHICAL DISTRIBUTION.

As before stated, this species was obtained at Hudson's Bay, and is supposed by RICHARDSON to exist on the Saskatchewan, thus giving it a considerable western range, should there not indeed prove to be a different species, which is, however, rather probable.

GENERAL REMARKS.

Until very recently there has been much confusion among writers in regard to the organization of the family of pouched-rats, which appear to be exclusively confined to the American continent—some supposing that the natural position of the pouch was that of a sac hanging suspended on each side of the throat, with the opening *within* the mouth.

For the probable origin of this error we refer our readers to the first volume of this work, p. 338, where we gave some remarks on the *Pseudostoma bursarius*, and this genus generally.

GENUS OVIBOS.—Blainville.

DENTAL FORMULA.

Incisive $\frac{0}{8}$; *Canine* $\frac{0—0}{0—0}$; *Molar* $\frac{6—6}{6—6} = 32$.

Body, low and compact; legs, short and covered with smooth short hairs; feet, hairy under the heel; forehead, broad and flat; no suborbital sinus; muzzle, blunt and covered with hair; horns, common to both sexes, in contact on the summit of the head, flat, broad, then tapering and bent down against the cheeks, with the points turned up; ears, short, and placed far back; eyes, small; tail, short.

Hair, very abundant, long, and woolly; size and form intermediate between the ox and the sheep; inhabits the northern or Arctic portions of North America.

The generic name is derived from two Latin words—*ovis*, sheep, and *bos*, ox.

There is only one known existing species of this genus, although *fossil* skulls have been found in Siberia, from which the name of *Ovibos pallentis* is given in systematic European works.

OVIBOS MOSCHATUS.—Gmel.

Musk-Ox.

PLATE CXI.—Males.

O. Fuscescente-niger, cornibus basi approximatis planis, latissimis, deorsum flexis, ad malas appressis apice extrorsum sursumque recurvis; mas magnitudine vaccæ biennis.

CHARACTERS.

Adult male, size of a small two year old cow; horns, united on the summit of the head, flat, broad, bent down against the cheeks, with the points turned up. Colour, brownish-black.

SYNONYMES.

Le Bœuf Musqué. M. Jeremie, Voyage au Nord, t. iii. p. 314.
    "   " Charlevoix, Nouv. France, tom. v. p. 194.

Musk-Ox. Drage, Voyage, vol. ii. p. 260.
"   Dobbs, Hudson's Bay, pp. 19, 25.
"   Ellis, Voyage, p. 232.
"   Pennant, Quadr., vol. i. p. 31.
"   "   Arctic Zoology, vol. i. p. 9.
"   Hearne's Journey, p. 137.
"   Parry's First Voyage, p. 257, plate.
"   "   Second Voyage, pp. 497, 503, 512 (specimen in British Museum).
Bos Moschatus. Gmel. Syst.
"   "   Capt. Sabine (Parry's First Voyage, Supplement, p. 189).
"   "   Mr. Sabine, Franklin's Journey, p. 668.
"   "   Richardson, Parry's Second Voyage, Appendix, p. 331.
Ovibos Moschatus. Richardson, Fauna Boreali Americana, p. 275.
Mataeh-Moostoos (Ugly Bison). Cree Indians.
Adgiddah-Yawseh (Little Bison). Chipewyans and Copper Indians.
Ooningmak. Esquimaux.
Ovibos Moschatus. Harlan, Fauna, p. 264.
Bos Moschatus—The Musk-Ox. Godman, Nat. Hist. vol. iii. p. 29.

DESCRIPTION.

Horns, very broad at base, covering the brow and crown of the head, touching each other for their entire basal breadth from the occipital to the frontal region : as the horns rise from their flatly-convex bases they become round and tapering, like those of a common cow, and curve downwards between the eye and the ear to a little below the eye, where they turn upwards and outwards (in a segment of a circle), to a little above the angle of the eye, ending with tolerably sharp points. The horns for half their length are rough, with small longitudinal splinters of unequal length, beyond which they are smooth and rather glossy, like those of a common bull.

Head, large and broad ; nose, very obtuse ; nostrils, oblong openings inclining towards each other downwards from above ; their inner margins naked ; united at their base. There is no other vestige of a muzzle ; the whole of the nose, and the lips, covered with a short coat of hairs ; there is no furrow on the upper lip.

The head, neck, and shoulders are covered with long bushy hair, and there is a quantity of long straight hair on the margins of the mouth and the sides of the lower jaw.

Eyes, moderately large, and the hair immediately around them shorter than on other parts of the cheeks ; ears, short, and scarcely visible through the surrounding long hair, which is more or less waved or crimped, and forms a sort of ruff back of the neck ; legs, short and thick, clothed with short hair unmixed with wool ; hoofs, flat, small in proportion to the size of the animal, and resembling those of the reindeer. The cow differs from the bull in having smaller horns (the bases of which, instead of touching each other, are separated by a hairy space), and in the hair on the throat and chest being shorter. The female is considerably smaller than the male.

COLOUR.

The general colour of the hair of the body is brown ; on the neck and between the shoulders it is of a grizzled hue, being dull light-brown, fading on the tips into brownish-white ; on the centre of the back it presents a soiled whitish colour, forming a mark which is aptly termed by Captain PARRY the saddle. The hips are dark-brown, and the sides, thighs, and belly, nearly black ; the short soft hairs on the nose and lips are whitish, with a tinge of reddish-brown ; legs, brownish-white ; tips of horns, and hoofs, black ; tail, dark brown.

DIMENSIONS.

	Feet.	Inches.
Length from nose to root of tail, about - -	5	6

HABITS.

For our description and account of the habits of this very peculiar animal we have resorted to other authors, never having ourselves had an opportunity of seeing it alive, and in fact knowing it only from the specimen in the British Museum, from which our figures were drawn, and which is the only one hitherto sent to Europe, so difficult is it to procure the animal and convey the skin, with the skull, leg bones, &c., in a tolerable state of preservation, from the barren lands of the northern portions of British America, where it is found, and where an almost perpetual winter and consequent scarcity of food make it very difficult to prevent the Indians, or white hunters either, from eating (we should say devouring) everything that can by any possibility serve to fill their empty stomachs—even skins, hoofs, and the most refuse parts of any animal they kill.

To give a better idea of the effects of hunger on man, at times, in these wild and desert countries, we will relate a case that happened to Dr. RICHARDSON while upon an expedition. One of his men, a half-breed and a bad fellow, it was discovered, had killed a companion with whom he had

Plate CXI

Drawn from Nature by J.W. Audubon.

Drawn on Stone by Wm E. Hitchcock

Musk Ox.

Lith. Printed & Cold by J.T. Bowen, Phil

been sent upon a short journey in the woods for intelligence, and had eaten a considerable portion of his miserable victim.

Dr. RICHARDSON, watching this monster from hour to hour, perceived that he was evidently preparing and awaiting an opportunity to kill him, possibly dreading the punishment he deserved for his horrible crime, and perhaps thinking the doctor's body would supply him with food till he could reach the settlements and escape :—anticipating his purpose, the doctor very properly shot him.

Sir JOHN relates an instance in which all his efforts to obtain a skin of the black-tailed deer were baffled by the appetites of his hunters, who ate up one they killed, hide and all. Even on the fertile prairies of more southern portions of our continent, starvation sometimes stares the hunter in the face. At one time a fine specimen of the mule deer (*Cervus macrotis*), shot for us on the prairies far up the Missouri river, was eaten by our men, who concealed the fact of their having killed the animal until some days afterwards.

Sir GEORGE SIMPSON, of the Hudson's Bay Fur Company, most kindly promised some years ago that he would if possible procure us a skin of the Musk-Ox, which he thought could be got within two years—taking one season to send the order for it to his men and another to get it and send the skin to England. We have not yet received this promised skin, and therefore feel sure that the hunters failed to obtain or to preserve one, for during the time that has elapsed we have received from the Hudson's Bay Company, through the kindness of Sir GEORGE, an Arctic fox, preserved in the flesh in rum, and a beautiful skin of the silver-gray fox, which were written for by Sir GEORGE at our request in 1845, at the same time that gentleman wrote for the skin of the Musk-Ox. We give an extract from Sir GEORGE's letter to us : " With reference to your application for skins of the Musk-Ox, I forwarded instructions on the subject to a gentleman stationed at the Hudson's Bay Company's post of Churchill, on Hudson's Bay, but the distance and difficulties of communication are so great that he will not receive my letter until next summer ; and he cannot possibly procure the specimens you require before next winter, nor can these be received in England before the month of October, 1847, and it is doubtful that they will be received even then, as those animals are scarce, and so extremely timid that a year might be lost before obtaining one."

Sir GEORGE SIMPSON was pleased to close this letter with a highly complimentary expression of the pleasure it would afford him to assist us in the completion of our work ; and among the difficulties and worrying accompaniments of such a publication as ours, it has been an unmixed gratification to have with us the sympathies and assistance of gentlemen like Sir GEORGE and many others, and of so powerful a corporation as the Hudson's Bay Fur Company.

Dr. RICHARDSON in a note explains a mistake made by PENNANT, who appears to have confounded the habitat of the Musk-Ox with that of the bison and states that our animal is found on the lands of the *Cris* or *Cristinaux* and *Assinibouls*, which are plains extending from the Red river of Lake Winnipeg to the Saskatchewan, on which tracts the buffalo is frequently found, but not the Musk-Ox.

The accounts of old writers, having reference to an animal found in New Mexico, which PENNANT refers to the Musk-Ox, may be based upon the existence of the Rocky Mountain sheep in that country, which having been imperfectly described, has led some authors to think the Musk-Ox was an inhabitant of so southern a locality.

" The country frequented by the Musk-Ox is mostly rocky, and destitute of wood except on the banks of the larger rivers, which are generally more or less thickly clothed with spruce trees. Their food is similar to that of the caribou—grass at one season and lichens at another ; and the contents of their paunch are eaten by the natives with the same relish that they devour the '*nerrooks*' of the reindeer. The droppings of the Musk-Ox take the form of round pellets, differing from those of the caribou only in their greater size.

" When this animal is fat, its flesh is well tasted, and resembles that of the caribou, but has a coarser grain. The flesh of the bulls is highly flavoured, and both bulls and cows, when lean, smell strongly of musk, their flesh at the same time being very dark and tough, and certainly far inferior to that of any other ruminating animal existing in North America.

" The carcase of a Musk-Ox weighs, exclusive of the offal, about three hundred weight, or nearly three times as much as a barren ground caribou, and twice as much as one of the woodland caribou.

" Notwithstanding the shortness of the legs of the Musk-Ox, it runs fast, and climbs hills or rocks with great ease. One, pursued on the banks of the Coppermine, scaled a lofty sand cliff, having so great an acclivity that we were obliged to crawl on hands and knees to follow it. Its foot-marks are very similar to those of the caribou, but are rather longer and narrower. These oxen assemble in herds of from twenty to thirty, rut about the end of August and beginning of September, and bring forth one calf about the latter end of May or beginning of June.

" HEARNE, from the circumstance of few bulls being seen, supposed that they kill each other in their contests for the cows. If the hunters keep themselves concealed when they fire upon a herd of Musk-Oxen, the poor animals mistake the noise for thunder, and, forming themselves into a group, crowd nearer and nearer together as their companions fall around them ; but should they discover their enemies by sight or by their sense of smell, which is very acute, the whole herd seek for safety by instant flight. The bulls, however, are very irascible, and particularly when wounded will often attack the hunter and endanger his life, unless he possess both activity and presence of mind. The Esquimaux, who are well accustomed to the pursuit of this animal, sometimes turn its irritable disposition to good account ; for an expert hunter having provoked a bull to attack him, wheels round it more quickly than it can turn, and by repeated stabs in the belly puts an end to its life. The wool of the Musk-Ox resembles that of the bison, but is perhaps finer, and would no doubt be highly useful in the arts if it could be procured in sufficient quantity."—*Richardson, F. B. A.*, p. 277.

"The Musk-Oxen killed on Melville Island during PARRY's visit, were very fat, and their flesh, especially the heart, although highly scented with musk, was considered very good food. When cut up it had all the appearance of beef for the market. HEARNE says that the flesh of the Musk-Ox does not at all resemble that of the bison, but is more like that of the moose, and the fat is of a clear white, tinged with light azure. The young cows and calves furnish a very palatable beef, but that of the old bulls is so intolerably musky as to be excessively disagreeable."—*Godman*, vol. iii. p. 35.

According to PARRY, this animal weighs about seven hundred pounds. The head and hide weigh about one hundred and thirty pounds. "The horns are employed for various purposes by the Indians and Esquimaux, especially for making cups and spoons. From the long hair growing on the neck and chest the Esquimaux make their musquito wigs, to defend their faces from those troublesome insects. The hide makes good soles for shoes and is much used for that purpose by the Indians."

GEOGRAPHICAL DISTRIBUTION.

The Musk-Ox resorts to the barren lands of America lying to the north of the 60th parallel of north latitude. HEARNE mentions that he once saw the tracks of one in the neighbourhood of Fort Churchill, lat. 59° ; and in his first journey to the north he saw many in the latitude of 61°. At present, according to what is said, they do not reach the shores of Hudson's Bay ; farther to the westward they are rarely seen in any number, lower than lat. 67°. RICHARDSON states that he had not heard of their being seen on the banks of Mackenzie's river to the southward of Great Bear lake. They range over the islands which lie to the north of the American continent as far as Melville Island, in latitude 75°, but they do not extend to Greenland, Lapland, or Spitzbergen. There is an extensive tract of barren country skirting the banks of the Mackenzie river, northwest of the Rocky Mountains, which also is inhabited by the Musk-Ox; it is not known in New Caledonia, on the banks of the Columbia, nor in any portion of the Rocky Mountains; nor does it cross over to the Asiatic shore: consequently it does not exist in any part of northern Asia or Siberia.

Captain PARRY noticed its appearance on Melville Island in the month of May; it must therefore be regarded as an animal the native home of which is within the Arctic Circle, the dwelling-place of the Esquimaux.

GENERAL REMARKS.

The Musk-Ox is remarkable amongst the animals of America, for never having had more than one specific appellation, whilst other species of much less interest·have been honoured with a long list of synonymes. JEREMIE appears to have given the first notice of it: he brought some of the wool to France, and had stockings made of it which were said to have been more beautiful than silk. The English voyagers of an early period gave some information respecting it, but PENNANT has the merit of being the first who systematically arranged and described it, from the skin of a specimen sent to England by HEARNE, the celebrated traveller. From its want of a naked muzzle and some other peculiarities, M. BLAINVILLE placed it in a genus intermediate (as its name denotes) between the sheep and the ox.

Plate CXII

Drawn from Nature by J.W. Audubon.

Drawn on stone by W.E. Hitchcock.

Californian Hare.

Lith. Printed & Col.d by J.T. Bowen, Phila.d

LEPUS CALIFORNICUS.—Gray.

Californian Hare.

PLATE CXII.

L. magnitudine L. glacialis, formâ L. timide ; supra flavescente-fuscus, subtus albus, flavo valdetinctus.

CHARACTERS.

Nearly the size of the polar hare ; dark brown on the back, light brownish-red on the neck ; lower parts deeply tinged with yellow.

SYNONYMES.

Lepus Californicus. Gray, Mag. Nat. Hist. 1837, vol. i., new series, p. 586.
   "   Richardsonii. Bach. Jour. Acad. Nat. Sci., vol. viii. p. 88.
   "   Bennettii. Gray, Zoology of the Voyage of H. M. S. Sulphur, Mamm., p. 35,
        pl. 14, 1843.

DESCRIPTION.

Head, small, and not elongated ; ears, very large, much longer than the head ; eyes, very large ; body, stout ; limbs, long and slender : fur, of moderate length ; tail, long and flat ; feet, rather small ; legs and feet, thickly clothed with short hairs nearly concealing the nails.

COLOUR.

The back, from the shoulder to the insertion of the tail, is strongly marked with black and rufous-brown, the hairs being pale plumbeous for two thirds of their length from the roots, then very pale brown, then black, then yellowish-brown, and tipped with black. Chest, sides of the body, and outer surface of limbs, more or less rufous. Abdomen, whitish tinged with buff ; upper surface of the tail blackish-brown, lower surface yellowish-white ; around the eye, pale buff ; back of the neck, grayish cinnamon colour ; legs and feet, cinnamon. The outer surface of the ears is longitudinally divided into two colours, the anterior portion or half being grizzled reddish-brown, becoming darker as it approaches the tip of the ear, the hairs being annulated with black and pale yellow ; the posterior portion dingy yellowish-white, growing lighter as it approaches the tip, until it blends with the black colour which terminates the upper half of the outside of the ear ; the interior edge of the ear is pale yellow, each hair slightly tipped with black ; one half of the inner surface of the ear is nearly naked, but covered with very delicate and short hairs, the other portion thinly clothed with hair gradually thickening towards the outer edge, where it is grizzly-brown ; edge of the ear for two thirds from the head, yellowish-white ; the remainder to the tip, soft velvety black. This black colour extends in a large patch on to the outer surface of the ear at the tip.

DIMENSIONS.

		Inches.	Lines.
Length from point of nose to root of tail,	- -	22	
" " eye to point of nose, - - -	-	2	1
Height of ear, posteriorly, - -	- -	5	10
Heel, to point of middle claw, - - -	-	4	8
Tail, including hair, - - - -	- -	3	3

HABITS.

The habits of all hares are much the same ; and this family is a general favourite for the beauty, timid gentleness, and fleetness its various species exhibit, although some of them are annoying to the gardener. In America, however, many species of Hare inhabit territories too far from cultivated fields or gardens for them to be able to nibble even at a cabbage plant.

Many pleasant evening hours have we passed, walking through forest-shaded roads in the last rays scattered here and there by the sinking sun, observing the playful "rabbits" leaping gracefully a few paces at a time, then stopping and looking about, ignorant of our proximity and unconscious of danger. But we are now to give the habits of the Californian Hare, for which take the following account of the animal as observed by J. W. AUDUBON :

"The Californian Hare appears to possess just brains enough to make him the greatest coward of all the tribe I have seen, for, once startled he is quite as wild as a deer, and equally heedless as to the course he takes, so that as he has not the keen sense of smell of the deer to warn him of danger in any direction, he sometimes makes a great fool of himself in his haste, and I have had these Hares run to within three feet of me, before I was seen, even where there was no cover but a sparse prairie grass."

"It was after toiling night and day through the sands of the Colorado desert, and resting afterwards at Vallecito and San Felipe, while marching along the streams through the rich fields of Santa Maria, that I saw the first Californian Hare. I knew him at sight: he showed no *white tail* as he ran, and looked almost black amongst the yellow broom-sedge as he divided it in his swift course. His legs seemed always under his body, for so quick was the movement that I could not see them extended, as in other Hares, from one bound to another; he seemed to alight on his feet perpendicularly at each leap, with a low-squatting springy touch to the earth, and putting his enormously long ears forward, and then back on his neck, and stretching out his head, appeared to fly over the undulating ridges of the prairie as a swallow skims for insects the surface of a sluggish river in summer."

Very few of these Hares were seen by J. W. AUDUBON's party until they had travelled some distance further north, and it was only after they had left the plains of the San Joaquin for the mines that they became a common animal, and in fact often their sole resource for the day's meat.

J. W. AUDUBON says that a single Hare of this species, with a little fat pork to fry it with, often lasted himself and a companion, as food when travelling, for two days. Nearly every miner has eaten of this fine Hare, which is well known in all the hilly portions of Upper California.

The Californian Hare brings forth about five young at a time, which are generally littered in the latter part of April or beginning of May. J. W. AUDUBON says: "I shot a female only a few days before her young would have been born: she had five beautiful little ones, the hair and feet perfect, and a white spot on the forehead of each was prominent. I never shot another afterwards, and was sad at the havoc I had committed."

We do not know whether this species breeds more than once in the year or not, but it probably does, as Mr. PEALE says: "A female killed on the twenty-fourth of September was still suckling her young."

The Californian Hare is more frequently met with in uplands, on mountain sides, and in bushy places, than in other situations. During the rainy season it was not seen by J. W. AUDUBON in low and wet grounds, although it doubtless resorts to them during the dry weather of summer.

Mr. PEALE says, these Hares "when running, carry the ears erect, and make three short and one long leap; and that the Indians catch them by setting hedges of thorny brush, with openings at intervals, in which they set snares, so constructed as to catch the Hares when passing, without the use of springes; the noose is made of a substance like hemp, very strong, and neatly twisted with cords."

GEOGRAPHICAL DISTRIBUTION.

This species was seen by J. W. AUDUBON during his journey from Texas to California; it was first met to the northward of the Colorado desert, and was quite abundant as the party approached the mining districts of California, where it was found as far north as the American fork; it was met with in the southern parts of Oregon by the United States Exploring Expedition. We are not informed whether it exists to the eastward of the Nevada range of mountains.

GENERAL REMARKS.

This Hare was first obtained by Mr. DOUGLAS, and sent with other animals from California to England. It was described by Mr. GRAY, and being, from its large size and rich colouring, one of the most conspicuous among the North American Hares, we regret that that eminent naturalist should have also (by some mistake) given it the name of *L. Bennettii*, and for ourselves we must plead guilty to having erroneously named it *L. Richardsonii*. The identity of this beautiful animal has been also somewhat obscured by Mr. PEALE, who confounded it with a species from the Cape of Good Hope, which bears the name of *Longicaudatus*, and was described in London.

Plate CXIII

Drawn from Nature by J.W. Audubon

Drawn on Stone by W.ᵐ E. Hitchcock

Esquimaux Dog

Lith. Printed & Col.ᵈ by J.T. Bowen Phil.

CANIS FAMILIARIS.—Linn. (Var. Borealis.—Desm.)

Esquimaux Dog.

PLATE CXIII.—Males.

C. magnitudine C. Terræ Novæ, capite parvo, auribus erectis, cauda comosa, cruribus pedibusque robustioribus, colore cinereo,.albo nigroque notato.

CHARACTERS.

About the size of the Newfoundland dog; head, small; ears, erect; tail, bushy; legs and feet, stout; general colour gray, varied with white and dark markings.

SYNONYMES.

Canis Familiaris, var. N. Borealis. Desm., Mamm., p. 194.
Esquimaux Dog. Captain Lyons, Private Journal, pp. 244, 332.
    "   "   Parry's Second Voyage, pp. 290, 358.
Canis Familiaris, var. A. Borealis—Esquimaux Dog. F. B. A., p. 75.

DESCRIPTION.

Head, rather small; ears, short and pointed; body, thick and well formed; eye, of moderate size; feet, clothed with thick short hair concealing the nails; tail, bushy, and longest at the end; hair, long, with thick wool beneath.

COLOUR.

Muzzle, black; inner portion of ears, blackish; top of nose, forehead, a space around the eyes, outer edges of ears, cheeks, belly, and legs, whitish; crown of the head, and back, nearly black; sides, thinly covered with long black, and some white, hairs; underneath there is a shorter dense coat of yellowish-gray woolly hair which is partly visible through these long hairs.

The tail, like the back, is clothed with black and white hairs, the latter greatly predominating, especially at the tip.

DIMENSIONS.

		Feet.	Inches
Length from point of nose to root of tail, -　-　-		4	3
"　"　of tail (vertebræ), -　-　-　-　-		1	2
"　"　including hair, 　-　-　-　-　-		1	5
Height of ear, inside, -　-　-　-　-　-　-			3
Width between the eyes, 　-　-　-　-　-　-			2¼
"　"　ears, 　-　-　-　-　-　-			4¼

HABITS.

So much has been written about the admirable qualities of the dog, that it would be quite useless for us to enter upon the subject; we shall also avoid the question of the origin of the various races, which in fact have been so intermixed that it would be an almost Quixotic task to endeavour to trace the genealogy of even the "noblest" of them. Those, however, that have, like the Esquimaux Dog, for centuries retained their general characters, and have not been exposed to any chance of "amalgamation" with other races, exhibit habits as well as forms and colours sufficiently permanent to warrant the naturalist in describing them, and in many cases their history is exceedingly interesting.

The Esquimaux Dogs are most useful animals to the savages of our Arctic regions, and when hitched to a sled many couples together, will travel with their master over the ice and snow at great speed for many miles without much fatigue, or draw heavy burthens to the huts of their owners. When on the coast of Labrador we had the following account of the mode in which these dogs subsist, from a man who had resided in that part of the world for upwards of ten years. During spring and summer they ramble along the shores, where they meet with abundance of dead fish, and in winter they eat the flesh of the seals which are killed and salted in the spring or late in the autumn when these animals return from the north. This man informed us also that when hard pushed he could relish the fare he thus provided for his Dogs just as much as they did themselves. We found several families inhabiting the coast of Labrador, all of whom depended entirely on their Dogs to convey them when visiting their neighbours, and some of whom had packs of at least forty of these animals. On some parts of the coast of Labrador the fish were so abundant during our visit that we could scoop them out of the edge of the water with a pocket-handkerchief: at such times the Esquimaux Dogs catch them, wading in and snapping at them with considerable dexterity as the surf retires; when caught they eat them at once while they are still alive.

We were informed that when these Dogs are on a journey, in winter, should they be overtaken by a severe snow-storm, and thereby prevented from reaching a settlement within the calculated time, and if the provisions intended for them in consequence give out, in their ravenous hunger they devour the driver, and even prey upon one another. Such cases were related to us, as well as others in which, by severe whipping and loud cries the Dogs were forced into a gallop and kept on the full run until some house was reached and the sleigh-driver saved.

These animals are taught to go in harness from the time they are quite young pups, being placed in a team along with well trained Dogs when only two or three months old, to gain experience and learn to obey their master, who wields a whip of twenty or thirty feet length of lash, with a short, heavy handle.

On a man approaching a house where they are kept, these Dogs sally forth with fierce barkings at the intruder, and it requires a bold heart to march up to them, as with their pointed ears and wiry hair they look like a pack of wild wolves. They are in fact very savage and ferocious at times, and require the strictest discipline to keep them in subjection.

Captain Lyon gives an interesting account of the Esquimaux Dog, part of which we shall here lay before you: " A walrus is frequently drawn along by three or four of these Dogs, and seals are sometimes carried home in the same manner, though I have in some instances seen a Dog bring home the greater part of a seal in panniers placed across his back. The latter mode of conveyance is often used in summer, and the Dogs also carry skins or furniture overland to the sledges when their masters are going on any expedition. It might be supposed that in so cold a climate these animals had peculiar periods of gestation, like the wild creatures ; but on the contrary, they bear young at every season of the year, the pups seldom exceeding five at a litter. Cold has very little effect on them ; for, although the dogs at the huts slept within the snow passages, mine at the ships had no shelter, but lay alongside, with the thermometer at 42° and 44° (below zero !) and with as little concern as if the weather had been mild. I found by several experiments, that three of my dogs could draw me on a sledge weighing 100 pounds at the rate of one mile in six minutes ; and as a proof of the strength of a well-grown Dog, my leader drew 196 pounds singly, and to the same distance, in eight minutes. At another time, seven of my Dogs ran a mile in four minutes, drawing a heavy sledge full of men. Afterwards, in carrying stores to the Fury, one mile distant, nine Dogs drew 1611 pounds in the space of nine minutes. My sledge was on runners neither shod nor iced ; but had the runners been iced, at least 40 pounds might have been added for each Dog,"

Captain Lyon had eleven of these Dogs, which he says "were large and even majestic looking animals; and an old one, of peculiar sagacity, was placed at their head by having a longer trace, so as to lead them through the safest and driest places." "The leader was instant in obeying the voice of the driver, who never beat, but repeatedly called to him by name. When the Dogs slackened their pace, the sight of a seal or a bird was sufficient to put them instantly to their full speed; and even though none of these might be seen on the ice, the cry of 'a seal!'—'a bear!'—'a bird!' &c., was enough to give play to the legs and voices of the whole pack. It was a beautiful sight to observe the two sledges racing at full speed to the same object, the Dogs and men in full cry, and the vehicles splashing through the holes of water with the velocity and spirit of rival stage-coaches. There is something of the spirit of professed whips in these wild races; for the young men delight in passing each other's sledge, and jockeying the hinder one by crossing the path. In passing on different routes the right hand is yielded, and should an inexperienced driver endeavour to take the left, he would have some difficulty in persuading his team to do so. The only unpleasant circumstance attending these races is, that a poor dog is sometimes entangled and thrown down, when the sledge, with perhaps a heavy load, is unavoidably drawn over his body.

"The driver sits on the fore part of the vehicle, from whence he jumps, when requisite, to pull it clear of any impediments which may lie in the way; and he also guides it by pressing either foot on the ice. The voice and long whip answer all the purposes of reins, and the Dogs can be made to turn a corner as dexterously as horses, though not in such an orderly manner, since they are constantly fighting; and I do not recollect to have seen one receive a flogging without instantly wreaking his passion on the ears of his neighbours. The cries of the men are not more melodious than those of the animals; and their wild looks and gestures, when animated, give them an appearance of devils driving wolves before them. Our Dogs had eaten nothing for forty-eight hours, and could not have gone over less than seventy miles of ground; yet they returned to all appearance as fresh and active as when they first set out."

These Dogs curl the tail over the hip in the manner of house dogs generally.

Our drawing was made from a fine living Dog in the Zoological Garden at London. Some have since been brought to New York alive by the ships fitted out and sent to the polar seas in search of the unfortunate Sir John Franklin and his party by Mr. Henry Grinnell, of that city.

GEOGRAPHICAL DISTRIBUTION.

This animal, as the name imports, is the constant companion of the Esquimaux, but extends much beyond the range of that tribe of Indians. since it is found not only at Labrador, but among various tribes of northern Indians, and was observed by travellers in the Arctic regions to the extreme north ; we are unacquainted with its western limits.

GENERAL REMARKS.

We have been induced, in our account of American animals, to give figures and descriptions of this peculiar variety of Dog, inasmuch as it appears to have been a permanent variety for ages, and is one of the most useful animals to the Indians residing in the polar regions. Whether it be an original native Dog, or derive its origin from the wolf, is a subject which we will not here discuss, farther than to state, in opposition to the views of Dr. RICHARDSON. that our figures do not represent these animals as very closely allied to the wolf ; on the contrary. their look of intelligence would indicate that they possess sagacity and aptitude for the service of man, equal at least to that of many favourite breeds of Dog. The fact also of their breeding at all seasons of the year, their manner of placing the tail in sport, and their general habits, give evidence of their being true Dogs and not wolves, the only difference between them and some other varieties consisting in their having erect pointed ears, which are peculiar to the Dogs of savage nations, and not altogether absent in some of our common breeds, as we have witnessed in the shepherd's Dog of Europe and some cur Dogs in America, erect ears of a similar character.

SPERMOPHILUS LATERALIS.—Say.

Say's Marmot-Squirrel, or Spermophile.

PLATE CXIV.

S. magnitudine Sciuri Hudsonici ; stria laterali flavescente alba nigro marginata.

CHARACTERS.

Size of Sciurus Hudsonicus ; *a yellowish-white stripe bordered with dark brownish-black on each flank.*

SYNONYMES.

Small gray Squirrel. Lewis and Clark, vol. iii. p. 35.
Sciurus Lateralis. Say, Long's Expedition, vol. ii. p. 46.
"   "   Harlan, Fauna Americana, p. 181.
Rocky Mountain Ground Squirrel. Godman, Nat. Hist., vol. ii. p. 144.
Arctomys (Spermophilus) Lateralis. Rich., Zool. Jour., vol. ii., No. 12, p. 519.
"   "   "   Say's Marmot. Rich., F. B. A., p. 174, pl. 13.

DESCRIPTION.

The body in form resembles the Spermophiles, with a slight approach to the Tamiæ ; head, rather large ; forehead, convex ; nose, obtuse and covered with short hairs, except a naked space around the nostrils ; incisors, flattened anteriorly ; mouth placed pretty far back ; whiskers, shorter than the head ; a few long black hairs over the eye and posterior part of the cheeks ; eyes, rather large ; ears, oval and somewhat conspicuous, appearing like the ears of most animals of this genus, with the exception that they seem as if trimmed or cut short ; they are thickly clothed on both surfaces with short hairs, and have a small doubling of the anterior margin to form a helix, which where it approaches the auditory canal is covered with longer hairs.

Legs, shorter and stouter than those of the squirrel family ; feet, shaped like those of the *Spermophili ;* claws, stronger, straighter, and better adapted for digging than those of the *Tamiæ ;* the thumb tubercle is far back, and has a small obtuse nail ; soles (of hind feet), naked to the heel, as are also the palms (of fore feet) and the under surface of the toes ; upper

Plate CXIV.

Drawn from Nature by J.W. Audubon.

Drawn on stone by W.E. Hitchcock

Say's Marmot Squirrel.

Lith Printed & Cold by J.T. Bowen, Philad.ª

surface of the feet, covered with short hairs which scarcely reach to the claws; tail depressed, slightly distichous, nearly linear, very slightly broadest towards the tip; there are no annulations in the hairs of the tail.

COLOUR.

Above, brownish-ash, intermixed with blackish, producing a hoary brownish-gray; there is no vestige of a dorsal line. A yellowish-white stripe appears on the neck, and running backwards along the sides, terminates at the hip; it is widest in the middle, being there three lines broad; and in some specimens it is faintly seen along the sides of the neck, reaching the ear; this white stripe is bounded above and below between the shoulder and the hip by a pretty broad border of brownish-black; top of the head and neck, tipped with ferruginous; the sides, all the ventral parts, inner surfaces of the legs, breast, and throat, yellowish-white, in parts tinged with brown.

Cheeks, and sides of the neck, chesnut-brown; ears, brown on their margins, paler near the base; a circle around the eye, upper lip, and chin, nearly white; nails, black; tail, black above, with an intermixture of brownish-white hairs, and bordered with white; the under surface is yellowish-brown, margined with black and brownish-white.

DIMENSIONS.

	Inches.	Lines
Length of head and body, - - - - - -	8	
" head, - - - - - - -	2	2
" tail (vertebræ), - - - - -	2	9
" " (including fur), - - - -	3	9
" middle fore claw, - - - -		4½
" palm and middle fore claw, - -		11
" sole and middle claw (of hind-foot), - -	1	6
Height of ear, - - - - - - -		4
Breadth of base of external ear, - - - -		5

HABITS.

This beautiful inhabitant of the wooded valleys of the Rocky Mountains was not seen by us on our journey up the Missouri river, although it is probably found within the district of country we traversed. We are therefore unable to give any personal information in regard to its habits, and we find but little in the works of others.

Mr. DRUMMOND obtained several specimens on the Rocky Mountains as far north as latitude 57°, and observed that it burrowed in the ground.

Mr. SAY did not give any account of its habits, and probably the specimen he described was brought into camp by the hunters attached to the expedition, without his ever having seen the animal alive.

All the Spermophiles that we have seen are lively, brisk, and playful, resembling the common ground-squirrels (*Tamias Lysteri*) in their general habits.

The Mexican women make pets of some of the species inhabiting that country, and they become very fond of their mistresses, running over their shoulders, and sometimes nestling in their bosoms, or the pockets of their gowns.

GEOGRAPHICAL DISTRIBUTION.

DRUMMOND obtained several specimens on the Rocky Mountains, in latitude 57°. LEWIS and CLARK state that it is common to every part of that range where wood abounds. We have not been able to determine the limits of its southern migrations, and have no information as to its existence in California.

GENERAL REMARKS.

This species was first observed by LEWIS and CLARK, but was named and described by Mr. SAY, who placed it among the ground-squirrels. Dr. RICHARDSON subsequently gave a very accurate description of it, and transferred it through *Arctomys* to the subgenus *Spermophilus*, although considering it intermediate between the nearly allied subgenera *Spermophilus* and *Tamias*, with respect to its claws and teeth.

It is, however, in reality a *Spermophilus* and not a *Tamias*, as can easily be seen from the form of the body, the shortness of the legs, shape of the feet, and more especially its strong and nearly straight nails. On the other hand, the longitudinal lines on the back, and the shape of the tail, indicate a slight approach to the *Tamiæ*.

At the close of this article we embrace the opportunity of adding another species to this interesting genus, the habitat of which is, however, we regret to say, so much involved in obscurity that we cannot with certainty, at present, add it to the list of our North American mammalia.

Shortly after the return of the United States Exploring Expedition under the command of Captain WILKES, we happened to meet several of the naturalists who had been attached to the expedition. Some one—we cannot now recollect the gentleman—presented us with this specimen, stating that he could not tell where it had been obtained ; the specimen has from that time remained in our collection without our having been able to gain any information in regard to its habitat, and without our learning that any other specimen has been procured, although we have anxiously sought to obtain farther intelligence on the subject.

This family is represented in the old world by few and peculiarly marked species, to none of which can we refer our animal, whilst on the other hand it bears in form, size, and markings, a strong connection with the American spermophiles, and will, as we are inclined to think, yet be found in some part of the western sea-coast regions of America.

We introduce it under the following name and description :

SPERMOPHILUS PEALEI.—Aud. and Bach.

S. Tamiâ Lysteri paullulum major ; striis albis quinque, cum quatuor fuscis alternantibus.

CHARACTERS.

A size larger than Tamias Lysteri ; *five white and four brown stripes.*

DESCRIPTION.

Head, smaller and shorter, and ears considerably longer and less abruptly terminated than in Say's *S. lateralis :* it is a little smaller than that species ; legs more slender, and tail longer, broader, and more distichous than in *S. lateralis ;* whiskers, long, a few of them extending beyond the ears.

On the fore feet there are four toes, without any vestige of a thumb or nail ; the claws are short and small, and are covered with hair ; palms, naked ; there are five toes on each hind foot ; the hair on the body is short and smooth, but is a little longer and also coarser on the under surface.

COLOUR.

A narrow white stripe rising on the back of the head runs along the centre of the back (or dorsal line) to the root of the tail ; another white stripe on each side originates behind the ear and runs along the upper part of the side, narrowing on the hips till it reaches the sides of the root of the tail ; a second white stripe on each side (lower than the last mentioned) runs from the shoulder to the hip, somewhat blended with a marked gray colour beneath it, which joins the colour of the under surface ; between these white stripes are four much broader : the two nearest the central white dorsal line are speckled light grayish-yellow and brown between the ears, gradually darkening into reddish on the centre of the back, and to brown near the tail ; the two outer brown stripes begin on the shoulder and run to the hips.

Forehead, speckled gray with a slight tinge of rufous towards the nose ; ears, thinly clothed with hair of a light gray on the outer surface and dull white within ; from the lower white stripe on each side, a grayish space extends between the shoulder and ham ; under the belly, inner sides of legs, throat, and chin, white ; the hams and shoulders are gray outside.

Whiskers, black ; teeth, orange ; nails, brown ; on the tail the hairs are yellowish-white from the roots, then black, then have a broader annulation of yellowish-white, then another of black, and are broadly tipped with white.

DIMENSIONS.

From point of nose to root of tail, - - - -	6½ inches.
Tail (vertebræ), - - - - - - - -	3½ "
" (to end of fur), - - - - - -	4½ "
Point of nose to ear, - - - - - - -	1¾ "
Height of ear, - - - - - - - -	⅝ "
Palm to end of middle nail, - - - - -	¾ "
Tarsus to longest nail on hind foot, - - - - -	¾ "

Plate CXV

Drawn from Nature by J.W.Audubon

On Stone by W.E.Hitchcock.

Lith.Printed & Col.d by J.T.Bowen, Phila.d

Yellow-Cheeked Meadow Mouse.

ARVICOLA XANTHOGNATHA.—Leach.

Yellow-cheeked Meadow-Mouse.

PLATE CXV.—Adult and Young.

A. Supra saturate fusca, subtus argenteo-cinereus, oculis circulo pallide luteo cinctis, genis flavis.

CHARACTERS.

Dark brown on the back; under parts, silvery grey; pale orange around the eyes; cheeks, yellow.

SYNONYMES.

Arvicola Xanthognatha. Leach, Zool. Miss., vol. i. p. 60, t. 26.
  "   "   Harlan, Fauna, p. 136.
  "   "   Godman, Nat. Hist., vol. ii. p. 65.
Campagnol aux joues fauves. Desm., Mamm., p. 282.
Arvicola Xanthognathus. Rich., Fauna Boreali Americana, p. 122.

DESCRIPTION.

Of the upper molars, the posterior one is the largest, and it has three grooves on its side; the two anterior have two grooves each, making in all ten ridges in the upper molar teeth on each side; of the lower molars, the anterior is the largest, and it has four grooves; the other two have each two.

Body, nearly cylindrical; legs, short; nose, obtuse; the lip is on a line with the incisors; ears, large, rounded, and hairy on both surfaces; whiskers, about the length of the head; tail, shorter than the head, well covered with hairs lying smoothly and coming to a point at the extremity; legs, rather stout, covered with short hair lying closely and smoothly; fore feet with naked palms; fore toes with a callosity protected by a very minute nail in place of a thumb; the first a little shorter than the third, second largest, and fourth shortest.

The toes are well covered with smooth hair above, and are naked below; the hair of the wrist projects a little over the palms; claws, small; hind feet with five toes, of which the three middle ones are nearly equal in length; the posterior part of the sole is covered with hair; soles of hind feet, narrower and longer than the palms of the fore feet; fur soft and fine, about four lines and a half long on the head, and nine on the posterior part of the back.

COLOUR.

The fur, from the roots to near the tips, is grayish-black; on the head and back the tips are yellowish-brown or black, the black pointed hairs being the longest; the colour resulting is a mixture of dark brown and black, without spots; sides, paler than the back; under parts, silvery bluish-gray.

Anterior to the shoulder, dark gray; there is a blackish-brown stripe on the centre of the nose; on each side of the nose a reddish-brown patch which extends to the orbit; around the eye, pale orange; whiskers, black; tail, brownish-black above, whitish beneath; feet, dark brown on the upper surface, whitish on the under.

DIMENSIONS.

	Inches.	Lines.
Length from point of nose to root of tail, - - -	8	
" of head, - - - - - - - -	1	10
" of tail, - - - - - - - -	1	6
Breadth of ear, - - - - - - - -		7
Hind foot, from heel to point of claw of middle toe, -		10

HABITS.

The descriptions of its habits given by the few writers who have referred with positive certainty to this species, are very meagre, but all the arvicolæ, with slight variations, are similar in habit; they live in low grounds, usually preferring meadows; burrow in the banks of ponds and near water-courses, feed on grasses and seeds, have a considerable number of young at a birth, are somewhat nocturnal, and make galleries of various lengths, which enable them to traverse the neighbourhood of their nestling places and procure the roots of grasses and plants.

This species, as is mentioned by RICHARDSON and other observers, makes its long galleries under the mossy turf, on the dry banks of lakes and rivers, and also in the woods; the specimens brought by us from Labrador were obtained from beneath large masses of moss growing on the rocks.

In some portions of the far north these hardy little animals are abundant : they were common in Labrador, and were easily captured by turning up some of the patches of moss, as just mentioned, when they were knocked over by the young men of our party.

We are told that this species has seven young at a time.

GEOGRAPHICAL DISTRIBUTION.

The original specimen described by LEACH, was obtained from Hudson's Bay : we procured several in Labrador.

Although supposed, by some writers, to exist within the limits of the United States, we have never been able to refer any species of Arvicola that has been discovered in our States or territories to this particular animal.

GENERAL REMARKS.

As before stated, LEACH described this Arvicola, and he also gave a very poor figure of it : SAY supposed it to exist on the banks of the Ohio, but we think he had in view a different species ; HARLAN appears not to have seen it, but gives the short description of LEACH, stating, however, that it exists in Pennsylvania and Ohio, which we presume was owing to his having mistaken for it some variety of WILSON's meadow-mouse (*A. Pennsylvanica*) ; GODMAN seems to have fallen into a similar mistake ; and the *Arvicola xanthognatha* of SABINE is evidently the *A. Pennsylvanica* of ORD.

Dr. DeKAY says it is found in various parts of the State of New York, but we have not been able to procure it, although we have sought for it for years ; and moreover we feel obliged to state that the description (which is a very unsatisfactory one), and the figure given in the "Zoology of the State of New York," refer to quite a different animal, probably one among the many varieties of *A. Pennsylvanica*.

We feel little hesitation in stating that this species does not exist in any part of the United States, but is exclusively a northern animal.

VULPES FULVUS.—DESM. (VAR. ARGENTATUS.—RICH.)

AMERICAN BLACK OR SILVER FOX.

PLATE CXVI.—FEMALE.

V. magnitudine V. fulvi, argenteo niger, cauda ad apicem alba.

CHARACTERS.

Size of the red fox (vulpes fulvus); *body, silvery black; tip of the tail, white.*

SYNONYMES.

RENARD NOIR OU BAHYNHA. Sagard Theodat., Canada, p. 744.
EUROPEAN FOX—var. A, black. Pennant, Arct. Zool., vol. i., p. 46.
RENARD NOIR OU ARGENTÈ. Geoffroy, Collect. du Museum.
GRIZZLED FOX. Hutchins, MSS.
RENARD ARGENTÈ. F. Cuvier, Mamm. Lith., 5 livr.
CANIS   GENTATUS. Desm., Mamm., p. 203.
     "   "   Sabine, Franklin's Journey, p. 657.
     "   "   Harlan, Fauna, p. 88.
     "   "   THE BLACK OR SILVER FOX. Godman, Nat. Hist., i. 274, plate.
     "   FULVUS, var. ARGENTATUS. Rich. BLACK OR SILVER FOX, F. B. A., p. 94.
BLACK FOX. DeKay, Nat. Hist. New York, p. 45.
TSCHERNOBURI. Russians.

DESCRIPTION.

Specimen from the Hon. Hudson's Bay Company.

Body, clothed with two kinds of hair; the longest, or outer hair, extends in some parts two inches beyond the under or shorter fur, especially on the neck, beneath the throat, behind the shoulders, along the flanks, and on the tail; this hair is soft, glossy, and finer than even that of the pine marten.

The under fur is unusually long and dense, measuring in some places two inches, and is exceedingly fine, feeling to the hand as soft as the finest sea-island cotton; this under fur surrounds the whole body even to the tail, on which it is a little coarser and has more the appearance of wool; it is shortest on the legs and forehead, and least dense on the belly; the hairs composing this fur, when viewed separately, exhibit a crimped or wavy appearance; on the ears and nose scarcely any long hairs are to be seen, these parts being thickly clothed with fur.

The soles of the feet are so thickly clothed with woolly hair that no callous spots are visible.

COLOUR.

The under fur is uniformly blackish-brown or chocolate; the long hairs are brown at their roots, then silver gray, and are broadly tipped with black; the hairs on the neck, and on a dorsal line extending to the root of the tail, are black, forming a broad black line at the neck, which narrows towards the tail.

Chin, throat, and whole under surface, brownish-black; a tuft of white hairs on the neck near the chest; another white tuft near the umbilicus; upper parts glossy silvery black; sides, sprinkled with many shining silvery white hairs, which produce a somewhat hoary appearance; tail, brownish-black to near the extremity, where it is broadly tipped with white.

DIMENSIONS.

	Feet.	Inches.
Nose to root of tail,	2	5
Length of tail,	1	7
Height of ear,		2¾
From nose to end of ear stretched back,		8½
" " eyes,		8⅜

HABITS.

Our account of the habits of this beautiful Fox will be perhaps less interesting to many than our description of its skin; for, as is well known, the Silver-gray Fox supplies one of the most valuable furs in the world, not only for the luxurious nobles of Russia and other parts of Europe, but for the old-fashioned, never-go-ahead Chinese, and other Eastern nations.

In the richness and beauty of its splendid fur the Silver-gray Fox surpasses the beaver or the sea-otter, and the skins are indeed so highly esteemed that the finest command extraordinary prices, and are always in demand.

The Silver-gray Fox is by no means abundant, and presents considerable variations both in colour and size. Some skins are brilliant black (with the exception of the end of the tail, which is invariably white); other specimens are bluish-gray, and many are tinged with a cinereous colour on the sides : it perhaps is most commonly obtained with parts of its fur hoary, the shiny black coat being thickly interspersed with white or silvery-blue tipped hairs.

According to Sir JOHN RICHARDSON, a greater number than four or five of these Foxes is seldom taken in a season at any one post in the fur countries, though the hunters no sooner find out the haunts of one than they use every art to catch it. From what he observed, Sir JOHN does not think this Fox displays more cunning in avoiding a snare than the red one, but the rarity of the animal, and the eagerness of the hunters to take it, make them think it peculiarly shy.

This animal appears to be as scarce in northern Europe as in America ; but we do not mean by this to be understood as considering the European Black Fox identical with ours.

The Black or Silver Fox is sometimes killed in Labrador, and on the Magdeleine Islands, and occasionally—very rarely—in the mountainous parts of Pennsylvania and the wilder portions of the northern counties of New York, where, however, PENNANT's marten is generally called the "Black Fox," by the hunters and farmers.

It gives us pleasure to render our thanks to the Hon. Hudson's Bay Company for a superb female Black or Silver-gray Fox which was procured for us, and sent to the Zoological Gardens in London alive, where J. W. AUDUBON was then making figures of some of the quadrupeds brought from the Arctic regions of our continent for this work. Having drawn this beautiful animal, which was at the time generously tendered us, but thinking it should remain in the Zoological Gardens, as we have no such establishment in America, J. W. AUDUBON declined the gift in favor of the Zoological Society, in whose interesting collection we hope it still exists. When shall we have a Zoological Garden in the United States?

This variety of the Fox does not differ in its propensities from the red Fox or the cross Fox, and its extraordinary cunning is often equalled by the tricks of these sly fellows.

The white tip at the end of the tail appears to be a characteristic of the Silver-gray Fox, and occurs in every specimen we have seen.

It is stated in MORTON's New England Canaan (p. 79), that the skin of the Black Fox was considered by the Indians, natives of that part of the colonies, as equivalent to forty beaver skins ; and when offered and accepted by their kings, it was looked upon as a sacred pledge of reconciliation.

The present species has been seen "mousing" in the meadows, near Ipswich, Massachusetts, as we were informed by the late WILLIAM OAKES,

Plate CXVI.

Drawn from Nature by J.W.Audubon.

On Stone by W.E.Hitchcock.

Lith.ᵈ Printed & Col.ᵈ by J.T.Bowen, Philad.ᵃ

American Black or Silver Fox.

who also wrote to us that "the common and cross Foxes were abundant about the White Mountains, and that they were most easily shot whilst scenting and following game, when their whole attention appears to be concentrated on that one object."

This Fox is occasionally seen in Nova Scotia, and a friend there informs us that some have been shot in his vicinity.

GEOGRAPHICAL DISTRIBUTION.

As this variety of the Red Fox chiefly occurs in the colder regions of our continent, we cannot set it down as a regular inhabitant of even the southern parts of the State of New York, nor any part of Pennsylvania or New Jersey.

The specimens which have been obtained in the two former States were killed at long intervals, and were, moreover, not of so fine a pelage or so beautiful a colour as those from more northern latitudes.

The skins sold to the American Fur Company are from the head waters of the Mississippi river, and the territories northwest of the Missouri, and are considered equal to the best.

GENERAL REMARKS.

The production of peculiar and permanent varieties in species of animals in a wild or natural state, is a subject of remarkable interest, although it cannot be explained on any data with which we are at present acquainted.

It is singular that in several species of red Foxes, widely removed from each other in their geographical ranges, the same peculiarities occur. The red Fox of Europe (*Canis vulpes*), a species differing from ours, produces no varieties in the southern and warmer parts of that continent, but is everywhere of the same reddish colour, yet in high northern latitudes, especially in mountainous regions, it exhibits not only the black, but the cross Fox varieties.

In the western portions of our continent the large red Fox of LEWIS and CLARK, which we described from a hunter's skin in our first volume (p. 54), and to which we have elsewhere given the name of *Vulpes Utah*, runs into similar varieties.

SCIURUS NIGRESCENS.—Bennett.

Dusky Squirrel.

PLATE CXVII.—Male.

S. Subniger, corpore griseo sparsim vario, lateribus flavo-fuscescentibus, cauda corpore multo longiore.

CHARACTERS.

*Prevailing colour dusky, slightly grizzled on the body with gray ; sides, .
dusky yellow ; tail much longer than the body.*

SYNONYMES.

Sciurus Nigrescens. Bennett, Proceedings of the Zoological Society.
      "   "   Bachman, Monog. Genus Sciurus, read before the Zool. Soc.,
                August 14, 1838.

DESCRIPTION.

In size this species is nearly equal to the cat-squirrel (*Sciurus cinereus*). Head, rather small ; ears, of moderate size, not tufted ; feet, robust ; tail, very long, and less distichous than in other squirrels, it presenting in the stuffed specimen a nearly cylindrical shape ; ears and feet, clothed with short hairs ; hairs of the body, short and close ; whiskers, about the length of the head.

COLOUR.

The prevailing colour on the back is grayish-black ; crown of the head, and legs, grayish ; sides of the neck, upper parts of the thighs, and rump, grizzled with pale yellow ; cheek, chin, throat, neck, breast, and whole of the under surface, including the inside of the legs, dingy gray ; fore parts, same colour as the back ; hairs of hinder parts of thighs, black ; hairs of the tail, black at the roots, then gray, then broadly banded with black, then broadly tipped with white ; feet, black.

The hairs on the toes are grizzled with white points ; whiskers, black ; hairs on the back, plumbeous—black from the roots for two thirds of their length, then gray, then black, and at the tips whitish-gray ; there are numerous strong black hairs interspersed over the body.

On Stone by W.E. Hitchcock.

Dusky Squirrel

Drawn from Nature by J.W.Audubon   Lith⁴ Printed & Col⁴ by J.T.Bowen, Philad⁴

DIMENSIONS.

	Inches.	Lines.
Length from point of nose to root of tail, - - -	12	4
" of tail to end of hair, - - - - -	15	4
" of tarsus (claws included), - - - -	2	7½
From tip of nose to ear, - - - - - -	2	2
Height of ear posteriorly, - - - - -		8½

HABITS.

The existence in North America of an unusual number of species of squirrels has been made known to our subscribers in the course of this publication. There are many closely allied, and many very beautiful species among them ; all are graceful and agile, and possess very similar habits.

The great number of these nut-eating animals in North America would be a proof (were any such wanting) that nature has been more bountiful to our country in distributing nut-bearing trees over the whole extent of our continent than to other parts of the globe, and this in connexion with the fact that so great a proportion of wood-land cannot be found in any other part of the world of similar extent, marks America as intended for a very dense population hereafter. In Europe there is only one well determined species of squirrel known, at present at least, although at some remote period there may have been more.

In regard to the peculiar habits of the Dusky Squirrel, we have nothing to say. It is one of the species which, being shot or procured by collectors of objects of natural history, and sent to Europe, have there been described by naturalists who, having the advantages of museums which contain specimens from every part of the globe, and the largest libraries in the world also to which they can refer, may sometimes discover new species with much less difficulty, but also less certainty, than the student of nature must encounter while seeking for knowledge in the woods.

But the naturalist who learns from books only, and describes from dried skins, is at best liable to mistakes. We have in fact always found that where young animals, or accidental varieties, have been described as new species, this has been the result of study in the museum or cabinet, not in the fields.

GEOGRAPHICAL DISTRIBUTION.

This species, of which, so far as we know, only one specimen exists in any museum or collection, is stated to have been procured in California.

We have not received any positive accounts of its occurrence there, but have no doubt it will be found, and its habits, as well as locality, determined ere long.

GENERAL REMARKS.

This Squirrel was described by Dr. BACHMAN from the original specimen in the museum of the Zoological Society of London, in his Monograph of the Genus *Sciurus*, published in the Proceedings of the Zoological Society, and in the Magazine of Natural History, new series, 1839, p. 113 ; and our figure was drawn from the same skin by J. W. AUDUBON.

Plate CXVIII

Drawn from Nature by J.W. Audubon.

On Stone by W.E. Hitchcock.

Lithd Printed & Cold by J.T. Bowen, Philad

Long-tailed Deer

CERVUS LEUCURUS.—Douglas.

Long-tailed Deer.

PLATE CXVIII.—Male.

C. Cervo Virginiano minor, capite atque dorso fulvis nigro mistis, malis lateribusque dilutioribus, gastræo albo.

CHARACTERS.

Smaller than the Virginian deer ; head and back, fawn-colour, mixed with black ; sides and cheeks, paler, white beneath.

SYNONYMES.

Roebuck. Dobbs, Hudson's Bay, p. 41, Ann. 1744.
Fallow, or Virginian Deer. Cook's Third Voyage, vol. ii. p. 292, Ann. 1778.
Long-tailed Jumping Deer. Umfreville, Hudson's Bay, p. 190, Ann. 1790.
Deer with Small Horns and Long tail (?) Gass, Journal, p. 55, Ann. 1808.
Long-tailed (?) Red Deer. Lewis and Clark, vol. ii. p. 41.
Small Deer of the Pacific. Idem, vol. ii. p. 342.
Jumping Deer. Hudson's Bay traders.
Chevreuil. Canadian Voyagers.
Mowitch. Indians west of the Rocky Mountains.

DESCRIPTION.

Form, elegant ; lachrymal opening, apparently only a small fold in the skin close to the eye ; limbs, slender ; hoofs, small and pointed ; tail, long in proportion to the size of the animal. Fur, dense and long ; a pendulous tuft of hairs on the belly between the thighs ; the glandular opening on the outside of the hind leg, small and oval in shape, the reversed hairs around it differing very little in colour from the rest of the leg. Hair, coarser than in the Virginian deer, and hoofs more delicate in shape.

COLOUR.

Head and back, rufous, mixed with black ; sides and cheeks, paler ; ears, above, dusky brown, inside edges, white ; there is a small black spot between the nostrils, and a white ring around the eyes. Chin and throat, yellowish-white ; tail, brownish-yellow above, inclining to rusty red near the tip, and cream white underneath and at the tip ; neck, brownish-yellow from the throat downwards ; under surface of the body, not so white as in the Virginian deer.

DIMENSIONS.

Young male in the Academy of Natural Sciences, Philadelphia.

	Feet.	Inches.
From point of nose to root of tail, - - -	4	2
Length of head, - - - - - - -		10½
End of nose to eye, - - - - - -		5½
Tail to end of hair, - - - - - -	1	1½
Height of ear posteriorly, - - - - -		5

Horns (two points about ¾ of an inch long, invisible without moving the surrounding hair).

Female presented by the Hudson's Bay Company to the museum of the Zoological Society.

	Feet.	Inches.
Length from point of nose to root of tail, - -	5	
" of head, - - - - - - -		11
" of tail (including fur), - - - -	1	1

HABITS.

In its general appearance this Deer greatly resembles the European roebuck, and seems to be formed for bounding along in the light and graceful manner of that animal. The species has been considered of doubtful authenticity, owing to the various lengths of tail exhibited by the common deer, many specimens of which we collected near the Rocky Mountains, not differing from *C. Virginianus* in any other particular, but with long tails, and for some time we did not feel inclined to give it a place in our work ; from which we have excluded a great many false species, published by others from young animals or mere varieties, and compared by us with specimens exhibiting all the markings and forms set down as characters by the authors alluded to. At one time we examined the tails of some common deer in Fulton market, New York, and found that the longest exceeded nineteen inches, while the average length does not go beyond nine. The different form of the light, springy animal described by Mr. DOUGLAS will, however, at once separate it from *C. Virginianus* on comparison.

Sir JOHN RICHARDSON says : "This animal, from the general resemblance it has in size, form, and habits, to the *Cervus capreolus* of Europe, has obtained the name of *Chevreuil* from the French Canadians, and of Roebuck from the Scottish Highlanders employed by the Hudson's Bay Company. These names occur in the works of several authors who have written on the fur countries, and UMFREVILLE gives a brief, but, as far as it goes, a correct description of it." "This species does not, on the east side of the Rocky Mountains, range farther north than latitude 54°, nor is it found in that parallel to the eastward of the 105th degree of longitude."

Mr. DOUGLAS speaks of it as "the most common deer of any in the districts adjoining the river Columbia, more especially in the fertile prairies of the Cowalidske and Multnomah rivers, within one hundred miles of the Pacific Ocean. It is also occasionally met with near the base of the Rocky Mountains on the same side of that ridge. Its favourite haunts are the coppices, composed of *Corylus, Rubus, Rosa,* and *Amelanchir,* on the declivities of the low hills or dry undulating grounds. Its gait is two ambling steps and a bound exceeding double the distance of the steps, which mode it does not depart from even when closely pursued. In running, the tail is erect, wagging from side to side, and from its unusual length is the most remarkable feature about the animal. The voice of the male calling the female is like the sound produced by blowing in the muzzle of a gun or in a hollow cane. The voice of the female calling the young is *mæ, mæ,* pronounced shortly. This is well imitated by the native tribes, with a stem of *Heracleum lanatum,* cut at a joint, leaving six inches of a tube : with this, aided by a head and horns of a full grown buck, which the hunter carries with him as a decoy, and which he moves backwards and forwards among the long grass, alternately feigning the voice with the tube, the unsuspecting animal is attracted within a few yards in the hope of finding its partner, when instantly springing up, the hunter plants an arrow in his object. The flesh is excellent when in good order, and remarkably tender and well flavoured." "They go in herds from November to April and May, when the female secretes herself to bring forth. The young are spotted with white until the middle of the first winter, when they change to the same colour as the most aged."

LEWIS and CLARK considered it the same animal as the common deer, with the exception of the length of the tail. They found it inhabiting "the Rocky Mountains, in the neighbourhood of the Chopunnish, and about the Columbia, and down the river as low as where the tide-water commences." These travellers in another passage observe that "the common Fallow Deer with long tails (our present species), though very poor, are better than the black-tailed fallow deer of the coast, from which they differ materially."

We did not see any Deer of this species on our journey up the Missouri, nor do we think it is to be found east of the Rocky Mountains. The Virginian deer, on the contrary, disappears to the north and west, as RICHARDSON says he has not been able to discover the true *Cervus Virginianus* within the district to which the Fauna Boreali Americana refers.

GEOGRAPHICAL DISTRIBUTION.

On the east side of the Rocky Mountains this species does not range beyond lat. 54°, nor to the eastward of 105° longitude. DOUGLAS states that it is the most common Deer of any in the districts adjoining the Columbia River, more especially in the fertile prairies of the Cowalidske and Multnomah rivers within one hundred miles of the Pacific Ocean. It is also occasionally met with near the base of the Rocky Mountains on the same side of that chain.

GENERAL REMARKS.

We have after some hesitation admitted this species, and as much has been said (although but little learned) of the western Long-tailed Deer since the days of LEWIS and CLARK, it is desirable that the species should be carefully investigated.

We overlooked the specimen of the Long-tailed Deer in the Zoological Museum, from which the description of RICHARDSON was taken, and for a long time we had no other knowledge of the species than the somewhat loose description of it by DOUGLAS, who, although an enthusiastic collector of plants and something of a botanist, was possessed of a very imperfect knowledge of birds or quadrupeds, and probably had never seen the *Cervus Virginianus*, our Virginian Deer.

We have given what we consider an excellent figure by J. W. AUDUBON, from the original specimen, and there is now in the Academy of Sciences at Philadelphia a young male which was procured some years since by the late Mr. J. K. TOWNSEND on the Columbia River.

Plate CIX

N° 24.

Drawn from Nature by J.W. Audubon.

Lith Printed & Cold by J.T. Bowen, Phila

On Stone by W.E. Hitchcock

Hudson's Bay Lemming

GENUS GEORYCHUS.—Illiger.

DENTAL FORMULA.

$$\textit{Incisive } \tfrac{2}{2}; \quad \textit{Canine } \tfrac{0-0}{0-0}; \quad \textit{Molar } \tfrac{3-3}{3-3} = 16.$$

This sub-genus in its dental formula is similar to *Arvicola ;* eyes, very small ; ears rising slightly above the auditory opening ; thumb, conspicuous ; nails on the fore feet fitted for digging ; tail, very short.

Natives of cold climates, burrow in the earth, feed on seeds, roots, and grasses.

Ten species are admitted by naturalists, two of which are in Europe, four in Asia, and four in America.

The generic name Georychus was given by ILLIGER, from Γεωρυχος, digging the earth.

———————

GEORYCHUS HUDSONIUS.—Forster.

Hudson's Bay Lemming.

PLATE CXIX.—Winter and Summer Pelage.

G. Auriculis nullis, maniculorum unguibus duobus intermediis, maximis, compressis, quasi duplicatis, per sulcum horizontalem divisis ; colore in æstate rufo-fusco, in hyeme albo.

CHARACTERS.

Earless : the two middle claws of the fore feet unusually large, compressed, their blunt extremity being rendered double by a deep transverse notch. Colour reddish-brown in summer, white in winter.

SYNONYMES.

Mus Hudsonius. Forster, Phil. Trans., vol. lxii. p. 379.
   "   "   Pallas, Glires, p. 208.
   "   "   Linn. Gmel. 137.
Hudson's Rat. Pennant, Quadrupeds, vol. ii. p. 201.
   "   "   "   Arctic Zoology, vol. i. p. 132.
Hare-tailed Mouse. Hearne's Journey, p. 387.

LEMMUS HUDSONIUS. Captain Sabine, Parry's Supplement, First Voyage, p. 185.
"   "   Mr. Sabine, Franklin's Journey, p. 661.
"   "   Dict. de Sci. Naturelles, tom. viii. p. 566.
"   "   Harlan, Fauna, p. 546.
ARVICOLA HUDSONIA. Rich., Parry's Second Voyage, Append., p. 308.
ARVICOLA (GEORYCHUS) HUDSONIUS—HUDSON'S BAY LEMMING. Rich., F. B. A., 132.
                    Species 107, British Museum.
HUDSON'S BAY LEMMING. Godman, Nat. Hist., vol. ii. p. 73.

DESCRIPTION.

Size of a mole ; body, thick and short ; head, short and rounded ; nose, very obtuse ; eyes, small ; no exterior ears ; legs, short and stout ; tail so short as to be only slightly visible beyond the fur of the hips ; fur very fine and long ; feet, clothed with long hairs ; four toes on the fore feet, with the rudiment of a thumb not armed with a nail ; the two middle toes are of equal length, and are each furnished with a disproportionately large claw, which is compressed, deep, very blunt at the extremity, and is there separated into two layers by a transverse furrow ; the outer and inner toes have curved sharp-pointed claws ; the upper layer is thinner, the lower one has a blunt rounded outline ; the latter has been described as an enlargement of the callosity which exists beneath the roots of the claws of the Lemmings and meadow-mice. The hind feet have five toes armed with slender curved claws.

In the females and young the subjacent production of the claws is less conspicuous.

COLOUR.

Winter specimen.

Whiskers, black ; the whole animal is white both on the upper and under surfaces, with black hairs interspersed along the line of the back and on the hips and sides, giving to those parts a grayish-brown tinge ; tail, white.

Summer specimen.

Dark brown and black on the dorsal aspect ; dark brown predominates on the crown of the head and dorsal line ; towards the sides the colour is lighter ; on the under parts of cheeks, the chest, and about the ears, bright nut colour prevails. The ventral aspect is grayish-white, more or less tinged with rust colour ; the tail is brown in summer, and white in winter ; although this species is distinctly white in winter, yet according to HEARNE the white colour never becomes so pure as that of the ermine.

DIMENSIONS.

							Inches.	Lines.
Length of head and body, -	-	-	-	-	-	5	4
" head, -	-	-	-	-	-	-	1	4
" tail, -	-	-	-	-	-	-		5
" middle fore claw,	-	-	-	-	-		4½

HABITS.

Our only acquaintance with this species is through the works of the old writers and the Fauna Boreali Americana, we having failed to meet with it at Labrador. The first specimen we saw of it was in the museum of the Royal College of Surgeons at Edinburgh. Our drawing was made from specimens in the British Museum. Dr. RICHARDSON did not meet with this Lemming in the interior of America, and thinks it has hitherto been found only near the sea.

"Its habits are still imperfectly known. In summer, according to HEARNE, it burrows under stones in dry ridges, and Captain SABINE informs us that in winter it resides in a nest of moss on the surface of the ground, rarely going abroad."—*Fauna Boreali Americana*, p. 132.

HEARNE states that this little species is very inoffensive, and so easily tamed that if taken even when full grown it will in a day or two be perfectly reconciled, very fond of being handled, and will creep of its own accord into its master's neck or bosom.

GEOGRAPHICAL DISTRIBUTION.

This species inhabits Labrador, Hudson's straits, and the coast from Churchill to the extremity of Melville peninsula, as well as the islands of the Polar seas visited by Captain PARRY.

GENERAL REMARKS.

This singular animal was originally described by FORSTER in the Philosophical Transactions. PALLAS received a number of skins from Labrador, one of which he sent to PENNANT, who described it in his History of the Quadrupeds and also in his Arctic Zoology. It was observed by both PARRY and FRANKLIN, and was described by RICHARDSON. A specimen was preserved in the Museum du Roi at Paris, and described in the Dict. des Sciences, and there is an excellent specimen in the British Museum.

GEORYCHUS HELVOLUS.—Rich.

TAWNY LEMMING.

PLATE CXX.—Fig. 1.

G. Pollice instructus, naso obtuso albido, capite fulvo nigroque vario, corpore supra fulvo, infra pallidiore, magnitudine G. Norvegici.

CHARACTERS.

Size of the Lapland Lemming ; nose, blunt and light coloured ; head, tawny black ; body, reddish-orange above, paler beneath ; feet, furnished with thumbs.

SYNONYMES.

ARVICOLA (LEMMUS) HELVOLUS. Richardson, Zool. Jour., No. 12, p. 517, 1828.
    "   (GEORYCHUS?) HELVOLUS. Rich., Fauna Boreali Americana, p. 128.

DESCRIPTION.

Body, stout ; head, oval ; nose, short, blunt, and nearly on a line with the incisors ; eyes, small ; ears, broad and not long—shorter than the fur, and clothed with hair near the edges ; tail, very short, clothed with stiff hairs, which are longest near the extremity, and converge to a point ; claws of both extremities much alike, greatly compressed, and sharp pointed ; the claws have an oblong narrow groove underneath.

The thumbs on the fore feet consist almost entirely of a thick, flat, squarish nail, resembling that of the Norway Lemming, and have, as in that species, an obliquely truncated summit ; in the Tawny Lemming, however, this summit presents two obscure points.

The fur on the body is about nine lines long ; that on the nose and extremities, very short.

COLOUR.

Body, reddish-orange, interspersed on the back and sides with a number of hairs longer than the fur, which are tipped with black ; on the upper parts of the head, around the eyes, and on the nape of the neck, the black hairs are more numerous, and the colour of those parts is mingled black and orange. Nose, grayish-brown ; sides of the face, pale orange ; margins of the upper lip, white ; tail, coloured like the body ; feet, brownish.

DIMENSIONS.

								Inches.	Lines.
Length of head and body,	-	-	-	-	-	-	-	4	6
" tail,	-	-	-	-	-	-	-		7
" head,	-	-	-	-	-	-	-	1	6
Hind feet to end of claw,	-	-	-	-	-	-			8
Fore feet and claws,	-	-	-	-	-	-			4½

HABITS.

Mr. DRUMMOND, who obtained this animal, procured no further information in regard to its habits than that it was found in Alpine swamps. It bears a strong resemblance to the Norway Lemming, and we may presume does not differ widely from that species in its habits, which it is said are migratory to a surprising extent, and about which some curious stories are related that we do not consider necessary to place in our work.

This Lemming is one of those animals we have never seen except the stuffed specimens. Our figure was drawn in London by J. W. AUDUBON from the original skin procured by Mr. DRUMMOND.

GEOGRAPHICAL DISTRIBUTION.

This animal was found in lat. 56°, in mountainous yet moist places, in the northwest. We have not heard of its existence in any other locality. but have no doubt it has a pretty extensive northern range.

GENERAL REMARKS.

The Lemmings have been arranged by authors, CUVIER, ILLIGER, and others, under a distinct subgenus—*Georychus*.

They are characterized chiefly by the shortness of the ears and tail, and large strong claws, remarkably well fitted for digging ; this subgenus, however, so nearly approaches the *Arvicolæ* in some of its species that it is difficult to decide in which genus they should really be placed.

GEORYCHUS TRIMUCRONATUS.—Rich.

PLATE CXX.—Figs. 2 and 3.

G. Auriculis vellere brevioribus, naso obtuso nigro, palmis tetradactylis, unguibus lanceolatis curvis, ungue pollicari lingulato, tricuspidato, corpore supra saturate castaneo. latere ferrugineo, subtus cinereo.

CHARACTERS.

Ears, somewhat shorter than the fur ; nose, blunt and black ; four claws on the fore feet of a lanceolate form, and a somewhat square thumb nail with three small points at the end ; body, dark chesnut above, reddish-orange or rust colour on the sides, gray beneath.

SYNONYMES.

Arvicola Trimucronatus. Rich, Parry's Second Voyage, Append., p. 309.
   "   (Georychus) Trimucronatus. Rich., Fauna Boreali Americana, p. 130.

DESCRIPTION.

In size a little inferior to the Hudson's Bay Lemming, or nearly equal to the Norwegian species ; head, flat and covered by moderately long fur ; ears, shorter than the fur, inclined backwards, and but thinly clothed with hair ; eyes, small. Upper lip, deeply cleft ; nose, obtuse, with a small naked but not pointed or projecting tip ; whiskers, numerous ; inside of the mouth, hairy, the hairs arising from projecting glandular folds ; upper incisors, presenting a conspicuous but shallow groove with an obliquely notched cutting edge ; there are three molar teeth on a side in each jaw. Fore legs, short ; feet, moderately large, and turned outwards like those of a turnspit.

The tail projects a few lines beyond the fur, and is clothed with stiff hairs converging to a point ; there are four toes on the fore feet, armed with moderate sized strong nails curved downwards and inclining outwards ; they are of an oblong form, convex above, not compressed, excavated underneath more broadly than the nails of any of the other American Lemmings, and have sharp edges fitted for scraping away the earth ; the thumb is almost entirely composed of a strong nail which has two slightly

Plate CXX

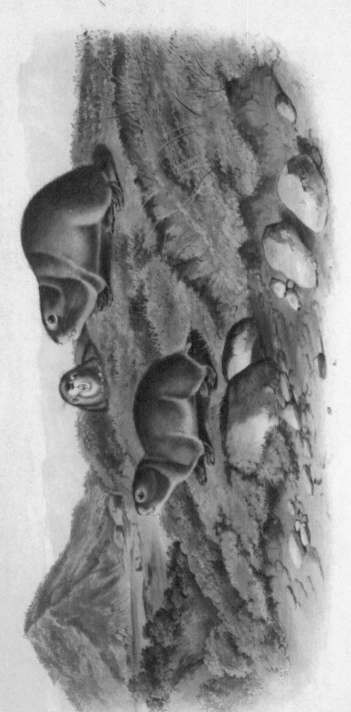

Drawn from Nature by J.W Audubon.

Lith. Printed & Col.d by J.T Bowen Phil.

Drawn on Stone by W.m J. Hitchcock.

Fig. 1. Tawny Lemming — Figs. 2 & 3. Back's Lemming

convex surfaces, a flat outline, and a truncated extremity from which three small points project; the palms are narrow: the posterior extremities are considerably longer than the fore legs and feet, the thighs and legs being tolerably distinct from the body: the sole is narrow, long, and somewhat oblique, having its inner edge turned a little forward: the toes are longer, and the claws as long but more slender than those of the fore feet, and they are much compressed.

In the Tawny Lemming the claws of both the fore and hind feet are compressed.

COLOUR.

Nose, deep black; whiskers, black at the roots, brownish or white at the tips, some entirely white; incisors, yellowish: head, back of the neck, and shoulders, mixed reddish-gray, formed from the mingling of yellowish and brown and black-tipped hairs: back, chesnut brown, with many of the long hairs tipped with black; sides, reddish-orange: belly, chin, and throat, gray, intermixed with many orange-coloured hairs.

The colouring of this animal very strongly resembles that of the Tawny Lemming, except that its nose is deep black, whilst that organ in the latter is pale.

Tail, dark brown above, grayish-white below; feet, dark yellowish-brown above, whiter beneath.

DIMENSIONS.

Male, killed at Fort Franklin.

		Inches.	Lines
Length of head and body, - - - - -	5	
" tail, - - - - - - -		6
" head, - - - - - - -	1	5
" ears, - - - - - - -		4
" whiskers, - - - - - -	1	3
" fur on the back, - - - - -		9
" palm and claw of middle toe, nearly -		6
" claw of middle toe, - - - -		2
" sole and middle claw of hind foot, -		9

Female 4¾ inches long.

HABITS.

This Lemming was found in the spring season at Great Bear Lake, by Sir JOHN FRANKLIN, burrowing under the thick mosses which cover a large portion of the ground in high northern latitudes.

As soon as the surface of the ground had thawed, the little animal was observed at work making his progress beneath, and actively engaged in hunting for food.

In the winter it travels under the snow in semi-cylindrical furrows, very neatly cut to the depth of two inches and a half in the mossy turf; these hollow ways intersect each other at various angles, but occasionally run to a considerable distance in a straight direction; from their smoothness it was evident that they were not merely worn by the feet, but actually cut by the teeth; their width is sufficient to allow the animal to pass with facility.

The food of this Lemming seems to consist entirely of vegetable matters; it inhabits woody spots.

A female killed on Point Lake, June 26, 1821, contained six young, fully formed, but destitute of hair.

GEOGRAPHICAL DISTRIBUTION.

This animal was discovered by Captain BACK on the borders of Point Lake, in latitude 65°, on Sir JOHN FRANKLIN'S first expedition. Mr. EDWARDS, the surgeon of the Fury, on Captain PARRY'S second expedition, brought a specimen from Igloolik, in latitude 69½°; and specimens were obtained on Sir JOHN FRANKLIN'S second expedition, on the shores of Great Bear Lake.

GENERAL REMARKS.

As we have been entirely unable to procure original information in regard to the habits of the two previously noticed and the present species of Lemming, we have largely quoted from the Fauna Boreali Americana, Sir JOHN RICHARDSON'S valuable work, from which also our descriptions of these curious animals are chiefly taken, although we have transposed the paragraphs in order to suit the general arrangement which we adopted for this work.

No animals belonging to this genus were observed by us during our researches through the country bordering on the shores of the upper Missouri and Yellow Stone rivers in 1843, and the family is very probably restricted to the neighbourhood of the Arctic Circle.

Plate CXXI

Drawn from Nature by J.W.Audubon.

Drawn on Stone by Wm. E. Hitchcock.

Lith. Printed & Cold by J.T. Bowen, Phil.

Arctic Fox.

VULPES LAGOPUS.—Linn.

Arctic Fox.

PLATE CXXI.—Winter and Summer Pelage.

V. Auriculis rotundatis brevibusque, margine inflexa ; collari post genas ; colore in æstate fusco, in hyeme albo.

CHARACTERS.

Ears, rounded, short, and folded at the edges ; cheeks with a ruff ; colour, in summer brown, in winter white.

SYNONYMES.

Pied Foxes. James's Voyage, Ann. 1633.
Canis Lagopus. Linn., Syst., vol. i. p. 59.
   "   "   Forster, Philos. Trans., lxii. p. 370.
Arctic Fox. Pennant's Arctic Zoology, vol. i. p. 42.
   "   "   Hearne's Journey, p. 363.
Greenland Dog. Pennant's Hist. Quadr., vol. i. p. 257 (?) a young individual.
Canis Lagopus. Captain Sabine, Parry's First Voyage, Supplement, 187.
   "   "   Mr. Sabine, Franklin's Journal, p. 658.
   "   "   Richardson, Parry's Second Voyage, Appendix, p. 299.
   "   "   Harlan, Fauna Americana, p. 92.
Isatis, or Arctic Fox. Godman's Nat. Hist., vol. i. p. 268.
Canis (Vulpes) Lagopus—Arctic Fox. Rich., Fauna Boreali Americana, p. 83.
Stone Fox. Auctorum.
Terreeanee-arioo. Esquimaux of Melville Peninsula.
Terienniak. Greenlanders.
Wappeeskeeshew-makkeeshew. Cree Indians.
Peszi. Russians.

DESCRIPTION.

Male in winter pelage.

Head, not as much pointed as in other species of Fox ; ears, rounded, and presenting somewhat the appearance of having been cropped ; hairs on the ears, shorter than on the neighbouring parts.

The cheeks are ornamented by a projecting ruff which extends from behind the ears quite around the lower part of the face, to which it gives a pleasing appearance; legs, rather long than otherwise, and muscular; feet, armed with pretty strong, long, compressed, and slightly arched claws; soles of the feet, covered with dense woolly hair; body covered with two kinds of hair, the longer thinly distributed and fine, the shorter a remarkably fine straight wool or dense fur; on the tail and lower parts of the body the long hairs are similar to those on the body, and the wool or fur like that of the finest wool of the merino sheep. The tail is thick, round, and bushy, and shorter than that of the red Fox.

The shoulders and thighs are protected by long fur, but the anterior parts of the legs are covered with short hair, the hind legs having the shortest and smoothest coat.

COLOUR.

In winter every part of this animal is white, except the tip of the nose the nails and eyes. Eyes, hazle; tip of nose, black; nails, brownish. The hairs of the animal are all white from the roots to the tips.

We have, however, seen specimens in which the colour was not pure white, but rather a bluish or brownish-gray tint at the roots on the back shoulders and outside of the thighs, but particularly on the neck and tail. The proportion of the fur so coloured varies with the season of the year as well as with different individuals of the species. Sometimes it is confined to a small space at the roots of the hair, whilst in other cases the dingy colour is so widely spread as to tarnish the customary whiteness of the whole skin.

At almost all times the short hair clothing the posterior surface and margin of the ears, is dark brownish-gray for half its length from the roots, so as to give a bluish or brownish tinge to view when the hairs are blown apart.

Summer pelage.

In the month of May, when the snow begins to disappear, the long white hairs and fur fall off, and are replaced by shorter hair, which is more or less coloured. A specimen killed at York factory on Hudson's Bay, in August, is described by Mr. SABINE as follows: "The head and chin are brown, having some fine white hairs scattered through the fur; the ears externally are coloured like the head; within they are white; a similar brown colour extends along the back to the tail, and from the back is continued down the outside of all the legs; the whole of the under parts, and the insides of the legs, are dingy white. The tail is brownish above, becoming whiter at the end, and is entirely white beneath."

DIMENSIONS.

Specimen obtained on the northeastern portion of the American continent by Captain PETTIGRU, and presented by him to the museum
of the Charleston College.

	Feet.	Inches.
From point of nose to root of tail, - - -	2	4
Length of tail (vertebræ), - - - - -	1	
" " (including fur), - - - -	1	2
" head, - - - - - - -		6
From point of nose to eye, - - - - -		2⅜
Height of ear anteriorly, - - - - -		2
From heel to point of middle claw, - - -		2⅛
Longest nail on the fore foot, - - - -		1⅒
" " hind foot, - - - -		⅜

Average weight about eight pounds, varying, according to Captain
LYON, from seven to nine and a half pounds when in good case.

HABITS.

From our description of the Arctic Fox, it will have been observed that
this animal is well adapted to endure the severest cold. In winter its feet
are thickly clothed with hair, even on the soles, which its movements on
the ice and snow do not wear away, as would be the case if it trod upon
the naked earth. These softly and thickly haired soles serve the double
purpose of preserving its feet from the effects of frost and enabling it
to run briskly and without slipping over the smooth icy tracts it must
traverse.

The Arctic Fox is a singular animal, presenting rather the appearance
of a little stumpy, round-eared cur, than that of the sharp and cunning-
looking Foxes of other species which are found in more temperate climes.
The character (for all animals have a character) and habits of this species
are in accordance with its appearance; it is comparatively unsuspicious
and gentle, and is less snappish and spiteful, even when first captured,
than any other Fox with which we are acquainted.

At times there is seen a variety of this Fox, which has been called the
Sooty Fox, but which is in all probability only the young, or at any rate
is not a permanent variety, and which does not turn white in winter,
although the species generally becomes white at that season. It is said
likewise that the white Arctic Foxes do not all assume a brown tint in
the summer. RICHARDSON says that only a majority of these animals

acquire the pure white dress even in winter ; many have a little duskiness on the nose, and others, probably young individuals, remain more or less coloured on the body all the year.   On the other hand, a pure white Arctic Fox is occasionally met with in the middle of summer, and forms the variety named *Kakkortak* by the Greenlanders.

Mr. WILLIAM MORTON, ship's steward of the Advance, one of Mr. HENRY GRINNELL's vessels sent in search of Sir JOHN FRANKLIN and his party, although not a naturalist, has furnished us with some account of this species.   He informs us that whilst the vessels (the Advance and Rescue) were in the ice, the men caught a good many Arctic Foxes in traps made of old empty barrels set with bait on the ice : they caught the same individuals in the same trap several times, their hunger or their want of caution leading them again into the barrel when only a short time released from captivity.

They were kept on board the vessels for some days, and afterwards let loose ; they did not always appear very anxious to make their escape from the ships, and those that had not been caught sometimes approached the vessels on the ice, where first one would appear, and after a while another, showing that several were in the neighbourhood.   They were occasionally observed on the rocks and snow on the land, but were not seen in packs like wolves ; they do not take to the water or attempt to swim.

These Foxes when they see a man do not appear to be frightened : they run a little way, and then sit down on their haunches like a dog, and face the enemy before running off entirely.   They are said to be good eating, the crews of the vessels having feasted on them, and are fat all the winter. They were occasionally seen following the polar bear to feed on his leavings, seals, flesh of any kind, or fish.

Those they captured were easily tamed, seldom attempting to bite even when first caught, and by wrapping a cloth around the hand some of them could be taken out of the barrel and held, not offering more resistance than a snap at the cloth.

Several beautiful skins of this animal were brought home by Dr. E. K. KANE, the accomplished surgeon of the expedition, and have since been presented by him to the Academy of Natural Sciences at Philadelphia.

Captain LYON, during two winters passed on Melville peninsula, studied with attention the manners of several of these animals.   He says : " The Arctic Fox is an extremely cleanly animal, being very careful not to dirt those places in which he eats or sleeps.   No unpleasant smell is to be perceived even in a male, which is a remarkable circumstance.   To come unawares on one of these creatures is, in my opinion, impossible, for even when in an apparently sound sleep they open their eyes at the slightest

noise which is made near them, although they pay no attention to sounds
when at a short distance. The general time of rest is during the daylight,
in which they appear listless and inactive ; but the night no sooner sets in
than all their faculties are awakened ; they commence their gambols, and
continue in unceasing and rapid motion until the morning. While hunting
for food, they are mute, but when in captivity or irritated, they utter a
short growl like that of a young puppy. It is a singular fact, that their
bark is so undulated as to give an idea that the animal is at a distance,
although at the very moment he lies at your feet.

" Although the rage of a newly caught Fox is quite ungovernable, yet it
very rarely happened that on two being put together they quarrelled. A
confinement of a few hours often sufficed to quiet these creatures ; and
some instances occurred of their being perfectly tame, although timid, from
the first moment of their captivity. On the other hand, there were some
which, after months of coaxing, never became more tractable. These we
suppose were old ones.

" Their first impulse on receiving food is to hide it as soon as possible,
even though suffering from hunger and having no fellow-prisoners of whose
honesty they are doubtful. In this case snow is of great assistance, as
being easily piled over their stores, and then forcibly pressed down by the
nose. I frequently observed my Dog-Fox, when no snow was attainable,
gather his chain into his mouth, and in that manner carefully coil it so as
to hide the meat. On moving away, satisfied with his operations, he of
course had drawn it after him again, and sometimes with great patience
repeated his labours five or six times, until in a passion he has been con-
strained to eat his food without its having been rendered luscious by previous concealment. Snow is the substitute for water to these creatures,
and on a large lump being given to them they break it in pieces with their
feet and roll on it with great delight. When the snow was slightly scattered on the decks, they did not lick it up as dogs are accustomed to do,
but by repeatedly pressing with their nose collected small lumps at its
extremity, and then drew it into the mouth with the assistance of the
tongue."

In another passage, Captain Lyon, alluding to the above-mentioned
Dog-Fox, says : " He was small and not perfectly white ; but his tameness
was so remarkable that I could not bear to kill him, but confined him on
deck in a small hutch, with a scope of chain. The little animal astonished
us very much by his extraordinary sagacity, for during the first day, finding
himself much tormented by being drawn out repeatedly by his chain,
he at length, whenever he retreated to his hut, took this carefully up
in his mouth, and drew it so completely after him that no one who

valued his fingers would endeavour to take hold of the end attached to the staple."

Richardson says that notwithstanding the degree of intelligence which the anecdotes related by Captain Lyon show them to possess, they are unlike the red Fox in being extremely unsuspicious; and instances are related of their standing by while the hunter is preparing the trap, and running headlong into it the moment he retires a few paces. Captain Lyon received fifteen from a single trap in four hours. The voice of the Arctic Fox is a kind of yelp, and when a man approaches their breeding places they put their heads out of their burrows and bark at him, allowing him to come so near that they may easily be shot.

They appear to have the power of decoying other animals within their reach, by imitating their voices. "While tenting, we observed a Fox prowling on a hill side, and heard him for several hours afterwards in different places, imitating the cry of a brentgoose." They feed on eggs, young birds, blubber, and carrion of any kind; but their principal food seems to be lemmings of different species.

Richardson thinks the "brown variety," as he calls it, the more common one in the neighbourhood of Behring's Straits. He states that they breed on the sea coast, and chiefly within the Arctic circle, forming burrows in sandy spots, not solitary like the red Fox, but in little villages, twenty or thirty burrows being constructed adjoining to each other. He saw one of these villages on Point Turnagain, in latitude 68½°. Towards the middle of winter, continues our author, they retire to the southward, evidently in search of food, keeping as much as possible on the coast, and going much farther to the southward in districts where the coast line is in the direction of their march. Captain Parry relates that the Arctic Foxes, which were previously numerous, began to retire from Melville peninsula in November, and that by January few remained. "Towards the centre of the continent, in latitude 65°, they are seen only in the winter, and then not in numbers; they are very scarce in latitude 61°, and at Carlton House, in latitude 53°, only two were seen in forty years. On the coast of Hudson's Bay, however, according to Hearne, they arrive at Churchill, in latitude 59°, about the middle of October, and afterwards receive reinforcements from the northward, until their numbers almost exceed credibility. Many are captured there by the hunters, and the greater part of the survivors cross the Churchill river as soon as it is frozen over, and continue their journey along the coast to Nelson and Severn rivers. In like manner they extend their migrations along the whole Labrador coast to the gulf of St. Lawrence. Most of those which travel far to the southward are destroyed by rapacious animals; and the few which survive to the spring breed in their new quar ters, instead of returning to the north. The colonies they found are however soon extirpated by their numerous enemies. A few breed at Churchill, and some young ones are occasionally seen in the vicinity of York factory. There are from three to five young ones in a litter."

The trap in which the Arctic Fox is taken by the Esquimaux, is described by authors as simple : it consists of a little hut built of stones, with a square opening on the top, over which some blades of whalebone are extended nearly across so as to form an apparently secure footing, although only fastened at one end, so that when the animal comes on to them to get the bait they bend downward and the Fox is precipitated into the hut below, which is deep enough to prevent his jumping out, the more especially because the whalebone immediately rises again to its position, and the bait being fastened thereto, several Foxes may be taken successively. Other traps are arranged so that a flat stone falls on the Fox when he by pulling at the bait disengages the trigger. These Foxes are also caught in traps made of ice (in which wolves are taken at times by the Esquimaux). These traps are thus described by Dr. RICHARDSON, and are certainly composed of the last material we, dwellers in more favoured lands, would think of for the purpose : "The Esquimaux wolf-trap is made of strong slabs of ice, long and narrow, so that a Fox can with difficulty turn himself in it, but a wolf must actually remain in the position in which he is taken. The door is a heavy portcullis of ice, sliding in two well-secured grooves of the same substance, and is kept up by a line, which, passing over the top of the trap, is carried through a hole at the farthest extremity ; to the end of the line is fastened a small hoop of whalebone, and to this any kind of flesh-bait is attached. From the slab which terminates the trap, a projection of ice or a peg of wood or bone points inwards near the bottom, and under this the hoop is slightly hooked ; the slightest pull at the bait liberates it, the door falls in an instant, and the wolf (or Fox) is speared where he lies."

In speaking of the *Sooty Fox*, which is only a variety of the present species, Dr. RICHARDSON says : "On one occasion during our late coasting voyage round the northern extremity of America, after cooking our supper on a sandy beach, we had retired to repose in the boats, anchored near the shore, when two Sooty Foxes came to the spot where the fire had been made, and carrying off all the scraps of meat that were left there, buried them in the sand above high water mark. We observed that they hid every piece in a separate place, and that they carried the largest pieces farthest off."

GEOGRAPHICAL DISTRIBUTION.

Arctic Foxes have been seen as far north on the American continent as man has ever proceeded. They are numerous on the shores of Hudson's Bay, north of Churchill, and exist also in Bhering's straits ; towards the centre of the continent in latitude 65°, they are seen only in the winter, and then not in numbers. They are very scarce in latitude 61°, and at Carlton house in latitude 53°, only two were seen in forty years. On the coast of Hudson's Bay, however, according to HEARNE, they arrive at Churchill, in latitude 59°, about the middle of October, and afterwards receive reinforcements from the northward. On the eastern coast of America they are found at Labrador, where they have been seen occasionally in considerable numbers ; a few have been also observed in the northern parts of Newfoundland, about latitude 52°.

On the eastern continent they are found in Siberia, and in all the Arctic regions.

GENERAL REMARKS.

We have had opportunities in the museums of London, Berlin, and more particularly at Dresden, of comparing specimens of this animal from both continents : we could not find the slightest difference, and have no hesitation in pronouncing them one and the same species.

Plate CXXII

Drawn from Nature by J.W.Audubon.

On Stone by W.E. Hitchcock.

Canada Otter

Lith. Printed & Col.d by J.T. Bowen, Phil.

LUTRA CANADENSIS.—Sabine. Var.
(Lataxina Mollis.—Gray.)

Canada Otter.

PLATE CXXII.—Male.

In our second volume (p. 12) we promised to give a figure of this variety of the Canada Otter, and in our remarks we noticed the publication of varieties of that animal as distinct species, by Gray, F. Cuvier, and Waterhouse.

Mr. Gray, we presume, thought that a larger and different species existed near Hudson's Bay, and named his specimen *Lataxina Mollis*, calling the animal the Great Northern Otter.

The figure now before you was published, notwithstanding our doubts as to the specific differences Mr. Gray thinks are observable between the Otters of Hudson's Bay and those of Canada and the United States, for the purpose of giving a correct drawing of the identical specimen named and described by that gentleman, in order that it might be seen that it is only a large variety of the common American Otter.

Besides giving a figure of Mr. Gray's Otter, we have examined Otters from very distant localities, having compared some taken near Montreal with one shot on the Hackensack river, New Jersey, several killed in South Carolina, one trapped in Texas, and one from California, and we are of opinion that, although differing in size and colour, the Otters of all these different localities are the same species, viz. *L. Canadensis*, the Canada Otter.

Besides the variations observable in the colour of the Otter, the fur of the more northern species is finer than in any of our southern specimens.

As already stated (vol. ii. p. 11) we have not had an opportunity of comparing specimens from Brazil with ours, and the description given by Ray of *Lutra Braziliensis* is so vague and unsatisfactory that we cannot state with confidence that his animal is identical with the North American species. We strongly suspect, however, that it is, in which case Ray's name, *L. Braziliensis*, should be substituted for *L. Canadensis*, to which we would add as synonymes *Lataxina Mollis* of Gray, and another supposed species by the same author, *Lutra Californica*.

We have nothing to add to the account of the habits of this animal given in our second volume (see p. 5).

VOL. III.—13

GENUS APLODONTIA.—Rich.

DENTAL FORMULA.

Incisive $\frac{2}{2}$; *Canine* $\frac{0-0}{0-0}$; *Molar* $\frac{5-5}{4-4} = 22.$

Incisors, very strong, flatly convex anteriorly, without grooves, narrower behind. Molars, simple, remarkably even on the crowns. The first in the upper jaw, small, cylindrical, and pointed, is placed within the anterior corner of the second one, and exists in the adult. The rest of the molars are perfectly simple in their structure, without roots, and have slightly concave crowns, which are merely bordered with enamel, without any transverse ridges or eminences. On the exterior side of the four posterior pairs of upper molars, and the inner side of all the lower ones, there is an acute vertical ridge extending the whole length of the tooth, formed by a sharp fold of enamel. When the molars are *in situ*, there is a wide semi-circular furrow between each pair of ridges, formed by the two adjoining teeth; the side of each tooth opposite the ridge is convexly semicircular. The second grinder in the upper jaw, and the first in the lower one, are a little larger than the more posterior ones, and the former has a projection of enamel at its anterior corner, producing a second though smaller vertical ridge, within which the first small molar is situated leaning towards it. There is a slight furrow on the exterior sides of the lower molars, most conspicuous in the first one.

Palate, narrow, bounded by perfectly parallel and straight rows of molars.

Head, flat and broad; nose, a little arched, thick, and obtuse. Lower jaw, thick and strong, with a large triangular process, concave behind, projecting at its posterior inferior angle further out than the zygomatic arch. The transverse diameter of the articulating surface of the condyle is greater than the longitudinal one. The jaw is altogether stronger than is usual in the *Rodentia*.

Cheek-pouches, none; eyes, very small; ears, short and rounded, approaching in form to the human ear, and thickly clothed with fur like that of a muskrat, but not so long or fine. Limbs, robust, short; feet, moderately long, with naked soles; five toes on all the feet, rather short but well separated; the thumb of the fore feet is considerably shorter than the other toes; claws, particularly the fore ones, very long, strong, much compressed, and but little curved.

Plate CXLIII.

Drawn from Nature by J.W. Audubon.

On Stone by W.E.Hitchcock

Lith. Printed & Col.ᵈ by J.T. Bowen, Philadᵃ

The Sewellel.

Tail, very short, concealed by the fur of the hips, mammæ six, the anterior pair situated between the fore legs.

Habits.—Form small societies, feeding on vegetable substances, and living in burrows.—*Richardson.*

There is only one species belonging to this genus known at present.

The name aplodontia is derived from απλοος, *aploos*, simple, and οδους, *odous*, a tooth.

APLODONTIA LEPORINA.—Rich.

The Sewellel.

PLATE CXXIII.—Male.—Natural Size.

A. Fuscescens, magnitudine Leporis Sylvatici, corpore brevi robusto, capite magno, cauda brevissima.

CHARACTERS.

Size of the gray rabbit (Lepus Sylvaticus). *Body, short and thick ; head, large ; tail, very short. Colour, brownish.*

SYNONYMES.

Sewellel. Lewis and Clark, vol. iii. p. 39.
Arctomys Rufa. Harlan, Fauna, p. 308.
    "   "   Griffith, Cuv. Animal Kingdom, vol. v. p. 245, species 636.
Aplodontia Leporina. Rich, Zool. Jour., No. 15, p. 335. January, 1829.
    "   '   —Sewellel. Rich, Fauna Boreali Americana, p. 211, pl. 18 c, figs. 7–14, cranium, &c.

DESCRIPTION.

Body, short, thick, and heavy, nearly reaching the ground ; legs, short ; head, large ; nose, thick and blunt, densely covered with hair to the nostrils, which are small and separated by a narrow furrowed septum concealed by the hair.

Mouth, rather small ; incisors, large and strong ; lips, thick, and clothed with rigid hairs ; a brush of white hair projects into the mouth from the upper lip near its union with the lower one ; whiskers, strong, and longer than the head ; a few stiff hairs over the eyes, on the cheeks, and on the outer sides of the fore-legs ; the eye is very small ; the external ear rises

rather far back, and is short and rounded ; it rises about four lines above
the auditory opening, has a small fold of the anterior part of its base
inwards, together with a narrow thick margin, representing a lobe.   There
are also folds and eminences in the cavity of the auricle ; the ear is clothed
on the outer surface with short and fine hairs, and on the inner, with
hairs a little longer ; tail, short, slender, and cylindrical, and almost
concealed by the hair of the rump ; legs, covered down to the wrists and
heels with short fur ; feet, shaped like those of the marmots ; palms and
under surfaces of the fore feet, naked ; there are three small callous
eminences at the roots of the toes, disposed as in the marmots, one of them
being common to the two middle toes, one proper to the third toe, and the
other to the little toe.

At the root of the thumb there is a large prominent callosity, and on the
opposite side of the palm another one nearly the same size ; the thumb is
of sufficient length to be used in grasping, and is terminated by a smooth
rounded nail ; claws, large and very much compressed, slightly arched
above, and nearly straight below ; hind feet, more slender than the fore
feet, and their claws one half smaller, rather more arched, and less compressed ; soles, longer than the palms, and naked to the heel ; they are
furnished with four callous eminences situated at the roots of the toes, and
two placed farther back, all more conspicuous than those on the hind feet
of the spermophiles of America.

The hair is soft, and somewhat resembles the finer fur of the muskrat ;
the under fur is soft, tolerably dense, and about half an inch long ; the
longer hairs are not sufficiently numerous to conceal the under fur.   The
hair on the feet only reaches to the roots of the claws, which are naked.

A specimen of a young Sewellel brought by DOUGLAS and examined by
RICHARDSON, in which the dentition was the same as in the adult, exhibited
a new set of molar teeth, which had destroyed the greater part of the substance of the old teeth, leaving merely a long process before and another
behind in each socket, resembling fangs.

COLOUR.

Incisors, yellow ; claws, horn colour ; general hue of the back, brownish,
the long scattered hairs being tipped with black ; belly, grayish, with
many of the long hairs tipped with white ; nose, nearly the colour of the
back ; lips, whitish ; in some specimens there is a spot of pure white on the
throat.

The hairs on the back, when blown aside, exhibit a grayish colour from
the roots to the tips, which are brown.

DIMENSIONS.

	Inches.	Lines.
Length of head and body, - - - - - -	14	
" tail, - - - - - - - -		6
Wrist joint to end of middle claw, - - - -	1	9
Middle claw, - - - - - - - -		6
Length of head, - - - - - - -	3	4

HABITS.

LEWIS and CLARK, who discovered this species during their journey across the Rocky Mountains to the Pacific, give us the following account of it :

"Sewellel is a name given by the natives to a small animal found in the timbered country on this coast. It is more abundant in the neighbourhood of the great falls and rapids of the Columbia than on the coast. The natives make great use of the skins of this animal in forming their robes, which they dress with the fur on, and attach them together with the sinews of the elk or deer. The skin when dressed is from fourteen to eighteen inches long, and from seven to nine in width : the tail is always separated from the skin by the natives when making their robes."

"This animal mounts a tree, and burrows in the ground, precisely like a squirrel. The ears are short, thin, and pointed, and covered with a fine short hair, of a uniform reddish-brown ; the bottom or the base of the long hairs, which exceed the fur but little in length, as well as the fur itself, are of a dark colour next to the skin for two thirds of the length of this animal ; the fur and hair are very fine, short, thickly set, and silky ; the ends of the fur and tip of the hair are of a reddish-brown, and that colour predominates in the usual appearance of the animal. Captain LEWIS offered considerable rewards to the Indians, but was never able to procure one of these animals alive."

Mr. DOUGLAS gave Dr. RICHARDSON an Indian blanket or robe, formed by sewing the skins of the Sewellel together. This robe contained twenty-seven skins, selected when the fur was in fine order. They are described by Dr. RICHARDSON as all having the long hairs so numerous as to hide the wool or down at their roots, and their points have a very high lustre. The doctor appears to think there were skins of two species of Sewellel in this robe. We did not hear of this animal ever being found to the east of the Rocky Mountains. Our figure was drawn from a fine specimen in London.

We are inclined to think from the form of the Sewellel that it is a great digger; but LEWIS' account of its mounting a tree seems to us to require some modification; the Maryland marmot, to which it is somewhat allied in form and in the shape of its claws, when hard pressed will mount a tree for a little distance to avoid the pursuit of a dog, but is very awkward and soon descends; we presume the climbing properties of the Sewellel can scarcely be greater than those of the marmot.

From the number of mammæ exhibited in the female, we conclude that it produces five or six young at a time, and from the nature of the animal, these are probably brought forth, like those of the marmots, in nests within their burrows.

GEOGRAPHICAL DISTRIBUTION.

This singular species has been observed on the western slopes of the Rocky Mountains, in the valleys and plains of the Columbia, at Nisqually, and at Puget's sound, where it is said to be a common animal. It has also been procured in California.

GENERAL REMARKS.

The history of this species, of which, however, little is known, is somewhat curious. LEWIS and CLARK appear to have been the only individuals who gave any notice of it until a very recent period, when DOUGLAS procured a specimen, and RICHARDSON gave a scientific account of the animal. The account LEWIS and CLARK gave dates back to 1804, and we have given the whole of their article above; these travellers, however, brought no specimens. After the journal of their adventurous expedition was published, RAFINESQUE ventured to give to the Sewellel the name of *Anysonix Rufa*, HARLAN named it *Arctomys Rufa*, and GRIFFITH introduced it into the animal kingdom under the same name; in 1829, RICHARDSON obtained a specimen, and the Sewellel was now for the first time examined by a naturalist. Believing that no one who had not seen or examined a species had a right to bestow a specific name, RICHARDSON rejected both the generic and specific names of previous writers, established for it a new genus, and gave it the name it now bears, and which it will doubtless preserve in our systems of Zoology.

There are two specimens of this animal in the Patent Office at Washington city, which were procured by the Exploring Expedition under command of Captain WILKES. We were recently politely refused permission to take them out of the glass case (in which they have for some time past remained) to examine their fur and measure them. We will not take the trouble to make any further remarks on this subject, as we have in a note at page 211 of our second volume mentioned the obstructions thrown in our way by the directors of the National Institute at Washington, the officers in charge of the collection informing us that by high authority the specimens were " tabooed."

PUTORIUS NIGRESCENS.—Aud. and Bach.

MOUNTAIN-BROOK MINK.

MOUNTAIN-BROOK MINK.

PLATE CXXIV (8vo. Ed.).—MALE.

P. Saturate fuscus, corpore minore quam in P. Visone, pedibus minus profunde palmatis, auriculis amplioribus et longioribus, vellere molliore et nitidiore quam in isto, dentibus longioribus in maxilla inferiore quam in superiore.

CHARACTERS.

Smaller than P. Vison ; *teeth in the under jaw larger than the corresponding teeth in the upper jaw ; feet, less deeply palmated than in* P. Vison ; *ears, broader and longer ; fur, softer and more glossy.   Colour, dark brownish-black.*

SYNONYME.

MOUNTAIN-MINK, of hunters.

DESCRIPTION.

In form, in dentition, and in the shape of the feet, this species bears a strong resemblance to a stout weasel ; the head is broad and depressed, and shorter and more blunt than the head of *Putorius Vison.*

Ears, large, oval, and slightly acute, covered on both surfaces with short fur ; legs, rather short and stout ; feet, small, and less webbed than in *P. Vison.* The callosities under the toes are more prominent than in that species, and the palms scarcely half as long. Whiskers, very numerous, springing from the sides of the face near the nose ; the body is covered with two kinds of hair, the under fur soft, and the long sparsely distributed hairs, coarse but smooth and glossy.

The toes are covered with short hairs almost concealing the nails, and the hairs between the toes leave only the tubercles or callosities on the under side of them visible.

COLOUR.

Fur, blackish-brown from the roots to the tips ; whiskers and ears, blackish-brown ; a patch on the chin, white ; under surface of body, a

Plate CXXIV

Drawn from Nature by J Woodhouse.

On Stone by Wᵐ E. Hitchcock.

Lith. Printed & Col₫ by J.T. Bowen, Phil.

Mountain Brook Mink.

shade lighter and redder than on the back ; tail, blackish-brown, except towards the tip, where it is black.

DIMENSIONS.

			Inches.	Lines.
Length of head and body, -	.	.	11	
" tail (to end of hair), -	.	.	7	
" " (vertebræ), -	.	.	6	
" palms of fore feet,	.	.	1	2
From tarsus to end of nail on hind foot, -	.	.	2	2
Height of ear externally, -	.	.		6

For convenient comparison we add the measurements of three common minks (*P. Vison*) killed in Carolina. One was very old and his teeth were much worn ; the other two were about eight months.

P. Vison, three specimens.

			Inches.	Inches.	Inches.
Lengths of body and head, respectively, -	.	20	17	19
" tail, "	.	8	6	7
" palms of fore feet, -	.	2		
" tarsus to longest nail,	.	8		

HABITS.

We were familiar with the manners and ways of this smaller mink in early life, and have frequently caught it in traps on the banks of a brook to which we resorted for the purpose of angling, and which in those days actually abounded with trout as well as with suckers and perch. On this sparkling stream, where we passed many an hour, the little black mink was the only species we observed. We found a nest of the animal under the roots of a large tree, where the young were brought forth, and we frequently noticed the old ones with fish in their mouths.

This species swim and dive swiftly and with apparent ease, but we most generally saw them on the ground, hunting as they stole along the winding banks of the stream, and following it high up into the hills towards its very source.

We remember seeing the young in the nest on two occasions ; in each case the nest contained four.

In early spring we have traced this species of Mink into the meadows, where it had been busily engaged in capturing the common meadow-mouse (*A. Pennsylvanica*), whilst the snow was yet on the ground.

Having one day detected one of these little Minks in an outhouse, closing the door immediately we captured it without its making any attempt either to get away or to defend itself. The frightened little marauder was probably conscious that it was in a prison from which there was no possible chance of escape.

The large species (*P. Vison*) appears to be more plentiful than the Mountain-brook Mink, and is found about mill-ponds and large rivers quite as frequently as on the borders of small streams.

The Mountain-brook Mink is quite as destructive to young poultry and to all the tenants of the farm-yard, when it happens to approach the precincts in which they may be thought to be safely ranging, as the larger species, or even the weasel.

GEOGRAPHICAL DISTRIBUTION.

We have observed this species in the mountains of the State of Pennsylvania, as well as in the northern part of the State of New York, in Vermont, and in Canada, but have not met with nor heard of it in Virginia or any of the Southern States, and consequently are inclined to regard it as a northern species.

It was not seen by us on the Missouri river, although it probably exists some distance to the west, in the latitude of the great lakes.

GENERAL REMARKS.

In our article on the common Mink (*Putorius Vison*, vol. i. p. 252) we referred to this smaller animal, but could not then find characters sufficient to separate the species.

Since that time, however, we have had abundant opportunities of comparing many specimens. We have seen some with their teeth much worn, and females which from the appearance of the teats had evidently suckled their young. They were all of the size and colour of the specimen above described, and we can no longer doubt that the latter is a distinct species from *P. Vison*.

The comparison in fact is not required to be made between these species, but between the present species and *P. lutreola* of Europe. We enjoyed opportunities of comparing *P. Vison* (the common and well known Mink) with the latter species in the museums of Berlin, Dresden, and London; but we had no opportunity of placing this little species by the side of the European.

We are inclined to believe, however, that the distinctive marks will be found in the small rounded feet and short tarsus of our present species, in its longer and rather more pointed ears, its shorter head, and longer lower incisors, together with a more general resemblance to our common weasel (*P. erminea*) in summer pelage.

SOREX PALUSTRIS.—Rich.

AMERICAN MARSH SHREW.

PLATE CXXV.—Males.

S. Mure musculo longior, cauda corporis fere longitudine, auriculis brevibus, pilosis, vellere absconditis, dorso canescente-nigro, ventro cinereo.

CHARACTERS.

Rather larger than the house mouse ; tail, nearly as long as the body ; short hairy ears, concealed by the fur ; back, somewhat hoary black ; belly, ash colour.

SYNONYMES.

Sorex Palustris. Rich., Zool. Jour., No. 12, April, 1828.
    "   "   " American Marsh Shrew. F. B. A.; p. 5.

DESCRIPTION.

Dental Formula.—Incisive $\frac{2}{2}$; Canine $\frac{4-4}{2-2}$; Molar $\frac{4-4}{3-3}$ = 30.
The two posterior lateral incisors are smaller than the two anterior ones on the same side, and the latter are a little longer than the posterior lobes of the intermediary incisors ; all the lateral incisors have small lobes on their inner sides. Muzzle, tolerably long, and pointed ; upper lip, bordered with rigid hairs ; tips of posterior hairs reaching beyond the ears ; the extremity of the muzzle, naked and bi-lobed ; eyes, small but visible ; ear, short and concealed by the fur, its margins folded in ; a heart-shaped lobe covering the auditory opening, and a transverse fold above it. The upper margins of the ears are clothed with thick tufts of fur. Tail rounded, and covered with hair, terminated by a small pencil of hair at the tip ; feet, clothed with rather short adpressed hairs, the hairs on the sides of the toes being arranged somewhat indistinctly in a parallel manner. The fur resembles that of the mole in softness, closeness, and lustre.

COLOUR.

The tips of the teeth have a shining chesnut-brown tint ; the body is black above, with a slight hoary appearance when turned to the light ; on

Plate LXV

N° 25

American Marsh Shrew

the ventral aspect ash coloured ; at the roots the hair is bluish-gray ; the outside of the thighs and upper surface of the tail correspond in colour with the back ; under surface of the tail, insides of thighs, and belly, greyish-white ; feet, paler than the back.

DIMENSIONS.

									Inches.	Lines.
Length from point of nose to root of tail,	-			-			-		3	6
" of tail,	-	-	-	-	-	-	-		2	7
" of head, -		-	-	-	-	-	-		1	2
" from nose to eye,	-	-	-	-	-	-				7
Height of ear,	-	-	-	-	-	-	-	-		3
Length of hind foot from heel to end of nails,						-		-		9

HABITS.

The habits of all Shrews (except those of the kind described by SHAKE-SPEARE) must necessarily be little known. These animals are so minute in the scale of quadrupeds that they will always be overlooked, unless sought after with great zeal, and even then it is often difficult to meet with or procure them. It may be said that it is only by chance that one is seen and taken now and then, even where they are known to exist. We have not seen more than five or six alive during several years, although dead ones have been found by us more frequently, and upon one occasion we found two that appeared to have recently died, lying close to each other. No wonder, then, that they may escape the observation of the most perse-vering student of nature, as their instinctive caution would, by causing them either to fly to some little hole or tuft of grass, or to remain still, when danger was near, render their discovery more than doubtful : or, if seen, it would be only for a moment. Not the least singular circumstance connected with the family of Shrews is the fact that they can exist in extremely cold climates, and move about in winter, when the snow covers the ground. In his article on *Sorex palustris* Dr. RICHARDSON says it " most probably lives in the summer on similar food with the Water Shrew, but I am at a loss to imagine how it procures a subsistence during the six months of the year in which the countries it inhabits are covered with snow. It frequents borders of lakes, and HEARNE tells us that it often takes up its abode in beaver houses."

We might easily make some probable speculations as to the manners and customs of the present species, but prefer not doing so farther than to say that it very likely feeds on seeds, insects, and on the carcases of any small birds or other animals it finds dead in the fields, that in winter it has a store of provision laid by, only coming to the snow-covered surface on fine days for the purpose of getting a little fresh air, and that from the number of tracks sometimes seen at one place we consider it partly gregarious in its habits.

Our drawing was made from a specimen in the British Museum at London.

GEOGRAPHICAL DISTRIBUTION.

The American Marsh Shrew, according to the writers who have seen it, exists in the northern parts of our continent from Hudson's Bay to the Coppermine river.

GENERAL REMARKS.

We are not aware that any author has referred to this animal, except Dr. RICHARDSON; the specimen from which our drawing was made was the original one from which Dr. RICHARDSON described, and we believe this species has never been hitherto figured.

GENUS RANGIFER.—Hamilton Smith.

DENTAL FORMULA.

Incisive $\frac{0}{8}$; *Canine* $\frac{1-1}{0-0}$; *Molar* $\frac{6-6}{6-6} = 34$.

Horns in both sexes, irregularly palmated, bifurcated, and rather long ; canine teeth in both sexes ; muzzle, small.

According to our opinion, two species of this genus exist—one in the old world (*Rangifer tarandus*), commonly called the Lapland Reindeer. and the Caribou (*Rangifer caribou*) and its varieties. the Reindeer of the American continent. Should, however, the varieties of the Reindeer found in different parts of the Arctic circle on *both* continents form one species only. then there is but one species in the genus known at present.

Fossil remains of a Reindeer of small size have been found near Etampes in France.

The generic name, *Rangifer*, is not of Latin origin, but has been formed from the old French term *Rangier* or *Ranger*, a Reindeer, probably through the later *Rangifère*.

RANGIFER CARIBOU.

Caribou or American Reindeer.

PLATE CXXVI.—Males. Fig. 1.—Summer Pelage. Fig. 2.—Winter

R. Magnitudine fere Elaphi Canadensis ; in æstate saturate fuscus, in hyeme cinereus ; vitta alba supra ungulas.

CHARACTERS.

Nearly the size of the American Elk (Elaphus Canadensis) ; *colour, deep brown in summer, grayish-ash in winter, a white fringe above the hoofs.*

SYNONYMES.

Genus Cervus. Linn., sectio Rangiferini.
Caribou, ou Asne Sauvage. Sagard Theodat. Canada, p. 751, Ann. 1636.
    "   La Hontan, t. i. p. 77, Ann. 1703.
    "   Charlevoix, Nouv. France. tom. v. p. 190.

REINDEER, or RAINDEER, Drage, Voy., vol. i. p. 25.
    "   Dobbs' Hudson's Bay, pp. 19, 22.
    "   Pennant's Arctic Zoology, vol. i. p. 22.
    "   Cartwright's Labrador, pp. 91, 112, 133.
    . "   Franklin's First Voyage, pp. 240, 245.
CERVUS TARANDUS. Harlan, Fauna, p. 232.
    "   "   Godman, Nat. Hist., vol. ii. p. 283.
    "   "   —REINDEER or CARIBOU, Rich., F. B. A., p. 238.
RANGIFER TARANDUS—REINDEER. DeKay, Nat. Hist. State of New York, p. 121.
ATTEHK. Cree Indians.
ETTHIN. Chippewyan Indians.
TOOKTOO. Esquimaux.
TUKTA. Greenlanders.
CARRÉ-BŒUF, or CARIBOU. French Canadians.

DESCRIPTION.

Young, about two years and a half old.

Larger and less graceful than the common American deer ; body, stout and heavy ; neck, short ; hoofs, thin, flattened, broad and spreading, excavated or concave beneath ; accessory hoofs, large but thin ; legs, stout ; no glandular opening and scarcely a perceptible inner tuft on the hind legs ; nose, somewhat like that of a cow, but fully covered with soft hairs of moderate length ; no beard, but on the under side of the neck a line of hairs about four inches in length which hang down in a longitudinal direction. Ears, small, short, and ovate, thickly clothed with hair on both surfaces ; horns, one foot three and a half inches in height, slender (one with two, and the other with one, prong) ; prongs, about five inches long.

Hair, soft and woolly underneath, the longer hairs like those of the antelope, crimped or waved, and about one to one and a half inches long.

COLOUR.

At the roots the hairs are whitish, then become brownish-gray, and at the tips are light dun gray, whiter on the neck than elsewhere ; nose, ears, outer surface of legs, and shoulder, brownish ; a slight shade of the same tint behind the fore legs.

Hoofs, black ; neck and throat, dull white ; a faint whitish patch on the sides of the shoulders ; forehead, brownish-white ; belly, white ; tail, white, with a slight shade of brown at the root and on the whole upper surface ; outside of legs, brown ; a band of white around all the legs adjoining the hoofs, and extending to the small secondary hoofs ; horns, yellowish-brown, worn whiter in places.

Plate CXXVI.

On Stone by Wᵐ E Hitchcock.

Drawn from Nature by J. W. Audubon.

Caribou or American Rein Deer.

Lith Printed & Col.ᵈ by J. T. Bowen, Phil.

There is a small patch of brown, faintly defined, around and behind the ears.

Description of the horns of another specimen.

The two main antlers are furnished with irregular and sharp points, and their extremity is pointed : some of these points are from six to eight inches long, but most of them are quite short : width between the horns on the skull, eight inches ; width of horns at the root, two inches and three quarters ; depth, one inch and three quarters : length of main horn, following the curve, three feet ; there is a palmated brow antler with four points, on one side, inclining downwards and inwards : on the opposite horn there are two points, but the antler is not palmated : immediately above the brow antlers there is a branch or prong on each horn about fourteen inches in length, terminating in three points : these prongs incline forward and inward. About half the length of the horn from the skull there is another prong on each about two inches long : beyond these prongs each horn continues about the same thickness, spreading outwards slightly to within a few inches of its extremity, where one diverges into five points and the other into six. The horns are but slightly channelled : they are dark yellow. Between the tips, where they approach each other, the horns are two feet apart, and at their greatest width two feet eight inches.

The female Caribou has horns as well as the male, but they are smaller.

DIMENSIONS.

Young—about two and a half years old.

	Feet.	Inches.
Length from nose to root of tail, - - - -	6	
" of tail (vertebræ), - - - - - -		4
" " (including hair), - - - - -		6½
Height of shoulder, - - - - - - -	3	6
Width between the eyes, - - - - - -		5½
From point of nose to lower canthus of eye, - -		9
" " to ear, - - - - - -	1	2
Height of ear posteriorly, - - - - - -		5

HABITS.

The Caribou, or American Reindeer, is one of the most important animals of the northern parts of America, and is almost as graceful in form as the elk (*Elaphus Canadensis*), to which it is nearly equal in size ; but it has never, we believe, been domesticated or trained to draw sledges in the manner of the Reindeer of the old world, although so nearly allied to that species that it has been by most authors considered identical with it.

Whilst separating the Caribou found in Maine and the States bordering on the St. Lawrence, and in Canada, Labrador, &c., from the Reindeer of Europe, we are inclined to think that the Reindeer found within the polar circle may be the European species, domiciled in that part of America, and that they sometimes migrate farther south than even Hudson's Bay. Sir JOHN RICHARDSON says the Reindeer or Caribou of North America " have indeed so great a general resemblance in appearance and manners to the Lapland Deer that they have always been considered to be the same species, without the fact having ever been completely established."—*Fauna Boreali Americana*, p. 238.

The greater size and weight of the Caribou found in Canada seem to have surprised Sir JOHN, but while he says in a note (p. 239), that " Mr. HENRY, when he mentions Caribou that weigh four hundred pounds, must have some other species of Deer in view," he has not done more than point out two varieties of Reindeer beside the one he considered identical with *Cervus tarandus* the European Reindeer, and to neither of these varieties can we with certainty refer the Caribou, our present animal. In the Fauna Boreali Americana (p. 241) one of these varieties—*C. tarandus, var. A. Arctica*, Barren-ground Caribou—is said to be so small that the bucks only weigh from ninety to one hundred and thirty pounds, exclusive of the offal, when in good condition ; the other variety—*C. tarandus, var. B. sylvestris*, Woodland Caribou (idem, p. 250)—is much larger than the Barren-ground Caribou, has smaller horns, and even when in good condition is vastly inferior as an article of food."

Leaving these supposed varieties where we found them—in doubt—we will proceed with an account of the habits of the Caribou detailed to us by Mr. JOHN MARTYN, Jr., of Quebec :

This species, that gentleman informed us, is not abundant near Quebec ; it is mostly found in the swamps, wherever these are well supplied with moss-covered dead trees and bushes ; the moss the animals prefer is a long and black species, and forms their chief subsistence during the winter months ; but towards spring these animals remove to the sides of the hills or mountains, and even ascend to their summits occasionally, feeding on the newly swollen buds of different shrubs. Like the moose deer they shed their antlers about this period, and renew them in the summer months.

The Caribou is famous for its swiftness, and has various gaits, walking trotting or galloping alike gracefully and rapidly. By many people these animals are in fact thought to be much fleeter animals than the moose, and they are said to take most extraordinary leaps.

When pursued the Caribou immediately makes for a swamp and follows the margin, taking at times to the water and again footing it over the firm ground, and sometimes turning towards the nearest mountain crosses it to another morass. If hard pressed by the hunters (who now and then follow up the chase for four or five days) the animal ascends to the loftiest peaks of the mountains for greater security. and the pursuit becomes very fatiguing and uncertain. Upon one occasion two men followed several Caribou for a whole week, when, completely tired out they gave up the chase, which was then continued by two other hunters who at last succeeded in killing a couple of the animals at long shot. Sometimes, however, fresh tracks are found and the Caribou is surprised whilst lying down or browsing, and shot on the spot. When the snow is not deep and the lakes are covered with ice only, the animal if closely pushed makes for one of them and runs over the ice so fast that it is unable to stop if struck with alarm at any object presenting itself in front, and it then suddenly squats down on its haunches and slides along in that ludicrous position until, the impetus being exhausted, it rises again and makes off in some other direction. When the Caribou takes to the ice the hunters always give up the chase.

Sometimes when the mouth and throat of a fresh killed Reindeer are examined they are found to be filled with a blackish looking mucus, resembling thin mud, but which appears to be only a portion of the partially decomposed black mosses upon which it fed, probably forced into the throat and mouth of the animal in its dying agonies.

We were informed that two wood-choppers, whilst felling trees at a distance from any settlement, saw a Caribou fawn approaching them which was so gentle that it allowed them to catch it, and one of the men took it up in his arms ; but suddenly the dam also made her appearance, and the men dropping the young one made after her in hopes of killing her with their axes. This object was of course soon abandoned, as a few bounds took the animal out of sight, and to their mortification they found that the fawn had escaped also during their short absence, and although they made diligent search for it, could not again be seen. At times, even the full grown Caribou appears to take but little heed of man.—A person descending a steep woody hill on a road towards a lake, saw several of them, which only turned aside far enough to let him pass, after which they came back to the road and proceeded at a slow pace up the hill. At another place a lad driving a cart was surprised to see five of these animals come into the road just before him, making a great noise through the woods. As soon as they got into the road they walked along quite leisurely, and on his cracking his whip only trotted a few paces and then resumed their walking.

When overtaken by dogs in chase, the Caribou stand at bay and show fight, and when thus brought to a stand will not pay much attention to the hunter, so that he can approach and shoot them with ease.

During our expeditions in Labrador we saw many trails of Reindeer through the deep and stiff moss; they are about as broad as a cowpath, and many times the fatigues of a long day's hunt over the sterile wilds of that country were lightened by following in these tracks or paths, instead of walking on the yielding moss.

We did not see any of these animals ourselves, but bought one from the Indians and enjoyed it very much, as we had had no fresh meat for nearly three months, except fishy ducks, a few curlews, and some willow-grouse.

We were informed that the Caribou are sometimes abundant on the island of Newfoundland, to which they cross on the ice from the mainland, and as the fishermen and French trappers at St. George's Bay told us, sometimes the herds stay so late in the spring that by the occasional early breaking up of the ice, they are prevented from leaving the island.

The horns of the Caribou run into various shapes, and are more or less palmated. The female of this species has also horns, which are not dropped until near the month of May. No two individuals of this species have the horns alike, nor do the horns of any grow into the same number of prongs, or resemble those of the last season. Notwithstanding this endless variety, there is always a specific character in the horns of this species (as well as in all our other deer), which will enable the close observer at once to recognise them.

"In the month of July," says Dr. RICHARDSON, "the Caribou sheds its winter covering, and acquires a short, smooth coat of hair, of a colour composed of clove brown, mingled with deep reddish and yellowish-browns, the under surface of the neck, the belly, and the inner sides of the extremities, remaining white in all seasons. The hair at first is fine and flexible, but as it lengthens it increases gradually in diameter at its roots, becoming at the same time white, soft, compressible, and brittle, like the hair of the moose deer. In the course of the winter the thickness of the hairs at their roots becomes so great that they are exceedingly close, and no longer lie down smoothly, but stand erect, and they are then so soft and tender below, that the flexible coloured points are easily rubbed off, and the fur appears white, especially on the flanks. This occurs in a smaller degree on the back; and on the under parts, the hair, although it acquires length, remains more flexible and slender at its roots, and is consequently not so subject to break. Towards the spring, when the Deer are tormented by the larvæ of the gad-fly making their way through the skin, they rub themselves against stones and rocks until all the coloured tops of the hair are worn off, and their fur appears to be entirely of a soiled white colour."

" The closeness of the hair of the Caribou, and the lightness of its skin, when properly dressed, render it the most appropriate article for winter clothing in the high latitudes. The skins of the young Deer make the best dresses, and they should be killed for that purpose in the month of August or September, as after the latter date the hair becomes too long and brittle. The prime parts of eight or ten Deer-skins make a complete suit of clothing for a grown person, which is so impervious to the cold that, with the addition of a blanket of the same material, any one so clothed may bivouack on the snow with safety, and even with comfort, in the most intense cold of an Arctic winter's night."

The same author gives the following habits of the variety he called " Arctica :" " The Barren-ground Caribou, which resort to the coast of the Arctic sea in summer, retire in winter to the woods lying between the sixty-third and the sixty-sixth degree of latitude, where they feed on the *usneæ, alectoriæ,* and other lichens, which hang from the trees, and on the long grass of the swamps. About the end of April, when the partial melting of the snow has softened the *cetrariæ, corniculariæ,* and *ceromyces,* which clothe the barren grounds like a carpet, they make short excursions from the woods, but return to them when the weather is frosty. In May the females proceed towards the sea-coast, and towards the end of June the males are in full march in the same direction. At that period the power of the sun has dried up the lichens on the barren grounds, and the Caribou frequent the moist pastures which cover the bottoms of the narrow valleys on the coasts and islands of the Arctic sea, where they graze on the sprouting carices and on the withered grass or hay of the preceding year, which is at that period still standing, and retaining part of its sap. Their spring journey is performed partly on the snow, and partly after the snow has disappeared, on the ice covering the rivers and lakes, which have in general a northerly direction. Soon after their arrival on the coast the females drop their young ; they commence their return to the south in September, and reach the vicinity of the woods towards the end of October, where they are joined by the males. This journey takes place after the snow has fallen, and they scrape it away with their feet to procure the lichens, which are then tender and pulpy, being preserved moist and unfrozen by the heat still remaining in the earth. Except in the rutting season, the bulk of the males and females live separately : the former retire deeper into the woods in winter, whilst herds of the pregnant does stay on the skirts of the barren grounds, and proceed to the coast very early in spring. Captain PARRY saw Deer on Melville peninsula as late as the 23d of September, and the females, with their fawns, made their first appearance on the 22d of April. The males in general do not go so far north as the females. On the coast of Hudson's Bay the Barren-ground Caribou migrate farther south than those on the Coppermine or Mackenzie rivers ; but none of them go to the southward of Churchill."

The Caribou becomes very fat at times, and is then an excellent article of food. As some particulars connected with its edible qualities are rather singular, we subjoin them from the same author : "When in condition there is a layer of fat deposited on the back and rump of the males to the depth of two or three inches or more, immediately under the skin, which is termed *depouillé* by the Canadian voyagers, and as an article of Indian trade, it is often of more value than all the remainder of the carcass. The *depouillé* is thickest at the commencement of the rutting season ; it then becomes of a red colour, and acquires a high flavour, and soon afterwards disappears. The females at that period are lean, but in the course of the winter they acquire a small *depouillé*, which is exhausted soon after they drop their young. The flesh of the Caribou is very tender, and its flavour when in season is, in my opinion, superior to that of the finest English venison, but when the animal is lean it is very insipid, the difference being greater between well fed and lean Caribou than any one can conceive who has not had an opportunity of judging. The lean meat fills the stomach but never satisfies the appetite, and scarcely serves to recruit the strength when exhausted by labour." "The Chepewyans, the Copper Indians, the Dog-Ribs and Hare Indians of Great Bear Lake, would be totally unable to inhabit their barren lands were it not for the immense herds of this Deer that exist there. Of the Caribou horns they form their fish-spears and hooks ; and previous to the introduction of European iron, ice-chisels and various other utensils were likewise made of them." "The hunter breaks the leg-bones of a recently slaughtered Deer, and while the marrow is still warm devours it with much relish. The kidneys and part of the intestines, particularly the thin folds of the third stomach or manyplies, are likewise occasionally eaten when raw, and the summits of the antlers, as long as they are soft, are also delicacies in a raw state. The colon or large gut is inverted, so as to preserve its fatty appendages, and is, when either roasted or boiled, one of the richest and most savoury morsels the country affords, either to the native or white resident. The remainder of the intestines, after being cleaned, are hung in the smoke for a few days and then broiled. The stomach and its contents, termed by the Esquimaux *nerrooks*, and by the Greenlanders *nerrokak* or *nerriookak*, are also eaten, and it would appear that the lichens and other vegetable matters on which the Caribou feeds are more easily digested by the human stomach when they have been mixed with the salivary and gastric juices of a ruminating animal. Many of the Indians and Canadian voyagers prefer this savoury mixture after it has undergone a degree of fermentation, or lain to season, as they term it, for a few days. The blood, if mixed in proper proportion with a strong decoction of fat meat, forms, after some nicety in the cooking, a rich soup, which is very palatable and highly nutritious, but very difficult of digestion. When all the soft parts of the animal are consumed the bones are pounded small, and a large quantity of marrow is extracted from them by boiling. This is used in making the better kinds of the mixture of dried meat and fat, which is named *pemmican*, and it is also preserved by the young men and females for anointing the hair and greasing the face on dress occasions. The tongue roasted, when fresh or when half dried, is a delicious morsel. When it is necessary to preserve the Caribou meat for use at a future period, it is cut into thin slices and dried over the smoke of a slow fire, and then pounded betwixt two stones. This pounded meat is very dry and husky if eaten alone, but when a quantity of the back-fat or *depouil'é* of the Deer is added to it, is one of the greatest treats that can be offered to a resident in the fur countries."

"The Caribou travel in herds, varying in number from eight or ten to two or three hundred, and their daily excursions are generally towards the quarter from whence the wind blows. The Indians kill them with the bow and arrow or gun, take them in snares, or spear them in crossing rivers or lakes. The Esquimaux also take them in traps ingeniously formed of ice or snow. Of all the Deer of North America they are the most easy of approach, and are slaughtered in the greatest numbers. A single family of Indians will sometimes destroy two or three hundred in a few weeks, and in many cases they are killed for the sake of their tongues alone."

Captain LYON's private journal contains some accounts of this species : "The Reindeer visits the polar regions at the latter end of May or the early part of June, and remains until late in September. On his first arrival he is thin and his flesh is tasteless, but the short summer is sufficient to fatten him to two or three inches on the haunches. When feeding on the level ground, an Esquimaux makes no attempt to approach him, but should a few rocks be near, the wary hunter feels secure of his prey. Behind one of these he cautiously creeps, and having laid himself very close, with his bow and arrow before him, imitates the bellow of the Deer when calling to each other. Sometimes, for more complete deception, the hunter wears his Deer-skin coat and hood so drawn over his head as to resemble, in a great measure, the unsuspecting animals he is enticing. Though the bellow proves a considerable attraction, yet if a man has great patience he may do without it, and may be equally certain that his prey will ultimately come to examine him, the reindeer being an inquisitive animal, and at the same time so silly that if he sees any suspicious object which is not actually chasing him, he will gradually and after many caperings, and forming repeated circles, approach nearer and nearer to it. The Esquimaux rarely shoot until the creature is within twelve paces, and I have frequently been told of their being killed at a much shorter distance. It is to be observed that the hunters never appear openly, but employ stratagem for their purpose ; thus, by patience and ingenuity, rendering their rudely formed bows and still worse arrows, as effective as the rifles of Europeans. When two men hunt in company they sometimes purposely show themselves to the Deer, and when his attention is fully engaged, walk slowly away from him, one before the other. The Deer follows, and when the hunters arrive near a stone, the foremost drops behind it and prepares his bow, while his companion continues walking steadily forward. This latter the Deer still follows unsuspectingly, and thus passes near the concealed man, who takes a deliberate aim and kills the animal. When the Deer assemble in herds there are particular passes which they invariably take, and on being driven to them are killed with arrows by the men, while the women with shouts drive them to the water. Here they swim with the ease and activity of water-dogs ; the people in kayaks chasing and easily spearing them ; the carcases float, and the hunter then presses forward and kills as many as he finds in his track. No springes or traps are used in the capture of these animals, as is practised to the southward, in consequence of the total absence of standing wood."

As presenting a striking illustration of the degree of cold prevailing in the Arctic regions, we may here mention that Dr. RICHARDSON describes a trap constructed by the Esquimaux to the southward of Chesterfield inlet, built of " compact snow." " The sides of the trap are built of slabs of that substance, cut as if for a snow house ; an inclined plane of snow leads to the entrance of the pit, which is about five feet deep, and of sufficient dimensions to contain two or three large Deer. The pit is covered with a large thin slab of snow, which the animal is enticed to tread upon by a quantity of the lichens on which it feeds being placed conspicuously on an eminence beyond the opening. The exterior of the trap is banked up with snow so as to resemble a natural hillock, and care is taken to render it so steep on all sides but one, that the Deer must pass over the mouth of the trap before it can reach the bait. The slab is sufficiently strong to bear the weight of a Deer until it has passed its middle, when it revolves on two short axles of wood, precipitates the Deer into the trap, and returns to its place again in consequence of the lower end being heavier than the other.

Throughout the whole line of coast frequented by the Esquimaux it is customary to see long lines of stones set on end, or of turfs piled up at intervals of about twenty yards, for the purpose of leading the Caribou to stations where they can be more easily approached. The natives find by experience that the animals in feeding imperceptibly take the line of direction of the objects thus placed before them, and the hunter can approach a herd that he sees from a distance, by gradually crawling from stone to stone, and remaining motionless when he sees any of the animals looking towards him. The whole of the barren grounds are intersected by Caribou paths, like sheep-tracks, which are of service to travellers at times in leading them to convenient crossing places of lakes or rivers."

The following account of a method of "impounding" Deer, resorted to by the Chepewyan Indians, is from HEARNE :

"When the Indians design to impound Deer, they look out for one of the paths in which a number of them have trod, and which is observed to be still frequented by them. When these paths cross a lake, a wide river, or a barren plain, they are found to be much the best for the purpose ; and if the path run through a cluster of woods, capable of affording materials for building the pound, it adds considerably to the commodiousness of the situation. The pound is built by making a strong fence with brushy trees, without observing any degree of regularity, and the work is continued to any extent, according to the pleasure of the builders. I have seen some that were not less than a mile round, and am informed that there are others still more extensive. The door or entrance of the pound is not larger than a common gate, and the inside is so crowded with small counter-hedges as very much to resemble a maze, in every opening of which they set a snare, made with thongs of parchment Deer-skins well twisted together, which are amazingly strong. One end of the snare is usually made fast to a growing pole ; but if no one of a sufficient size can be found near the place where the snare is set, a loose pole is substituted in its room, which is always of such size and length that a Deer cannot drag it far before it gets entangled among the other woods, which are all left standing, except what is found necessary for making the fence, hedges, &c. The pound being thus prepared, a row of small brush-wood is stuck up in the snow on each side of the door or entrance, and these hedge-rows are continued along the open part of the lake, river, or plain, where neither stick nor stump besides is to be seen, which makes them the more distinctly observed. These poles or brushwood are generally placed at the distance of fifteen or twenty yards from each other, and ranged in such a manner as to form two sides of a long acute angle, growing gradually wider in proportion to the distance they extend from the pound, which sometimes is not less than two or three miles, while the Deer's path is exactly along the middle, between the two rows of brushwood. Indians employed on this service always pitch their tents on or near to an eminence that affords a commanding prospect of the path leading to the pound, and when they see any Deer going that way, men, women, and children walk along the lake or river side under cover of the woods, till they get behind them, then step forth to open view, and proceed towards the pound in form of a crescent. The poor timorous Deer, finding themselves pursued, and at the same time taking the two rows of brushy poles to be two ranks of people stationed to prevent their passing on either side, run straight forward in the path till they get into the pound. The Indians then close in, and block up the entrance with some brushy trees that have been cut down and lie at hand for that purpose. The Deer being thus enclosed, the women and children walk round the pound to prevent them from jumping over or breaking through the fence, while the men are employed spearing such as are entangled in the snares, and shooting with bows and arrows those which remain loose in the pound. This method of hunting, if it deserve the name, is sometimes so successful that many families subsist by it without having occasion to move their tents above once or twice during the course of a whole winter; and when the spring advances, both the Deer and the Indians draw out to the eastward on the ground which is entirely barren, or at least which is called so in these parts, as it neither produces trees nor shrubs of any kind, so that moss and some little grass is all the herbage which is to be found on it."

With the following extract from the Fauna Boreali Americana, our readers may perhaps be amused : " The Dog-rib Indians have a mode of killing these animals, which, though simple, is very successful. It was thus described by Mr. WENTZEL, who resided long amongst that people : The hunters go in pairs, the foremost man carrying in one hand the horns and part of the skin of the head of a Deer, and in the other a small bundle of twigs, against which he, from time to time, rubs the horns, imitating the gestures peculiar to the animal. His comrade follows, treading exactly in his footsteps, and holding the guns of both in a horizontal position, so that the muzzles project under the arms of him who carries the head. Both hunters have a fillet of white skin round their foreheads, and the foremost has a strip of the same around his wrists. They approach the herd by degrees, raising their legs very slowly but setting them down somewhat suddenly after the manner of a Deer, and always taking care to lift their right or left feet simultaneously. If any of the herd leave off feeding to gaze upon this extraordinary phenomenon it instantly stops, and the head begins to play its part by licking its shoulders and performing other necessary movements. In this way the hunters attain the very centre of the herd without exciting suspicion, and have leisure to single out the fattest. The hindmost man then pushes forward his comrade's gun, the head is dropt, and they both fire nearly at the same instant. The Deer scamper off, the hunters trot after them ; in a short time the poor animals halt to ascertain the cause of their terror, their foes stop at the same moment, and having loaded as they ran, greet the gazers with a second fatal discharge. The consternation of the Deer increases : they run to and fro in the utmost confusion, and sometimes a great part of the herd is destroyed within the space of a few hundred yards."

We do not exactly comprehend how the acute sense of smell peculiar to the Reindeer should be useless in such cases, and should think the Deer could only be approached by keeping to the leeward of them, and that it would be a very difficult matter, even with the ingenious disguise adopted by the "Dog-Ribs," to get into the centre of a herd and leisurely single out the fattest.

Dr. RICHARDSON considers the variety he calls the woodland Caribou as much larger than the other, and says it has smaller horns, and is even when in good condition vastly inferior as an article of food. "The proper country of this Deer," he continues, "is a stripe of low primitive rocks, well clothed with wood, about one hundred miles wide, and extending at the distance of eighty or a hundred miles from the shores of Hudson's Bay, from Athapescow Lake to Lake Superior. Contrary to the practice of the barren-ground Caribou, the woodland variety travels to the southward in the spring. They cross the Nelson and Severn rivers in immense herds in the month of May, pass the summer on the low marshy shores of James' Bay, and return to the northward, and at the same time retire more inland in the month of September."

GEOGRAPHICAL DISTRIBUTION.

This species exists in Newfoundland and Labrador, extends westward across the American continent, and is mentioned both by PENNANT and LANGSDORFF as inhabiting the Fox or Aleutian Islands.

It is not found so far to the southward on the Pacific as on the Atlantic coast, and is not found on the Rocky Mountains, within the limits of the United States. According to PENNANT there are no Reindeer on the islands that lie between Asia and America. It is somewhat difficult to assign limits to the range of the Caribou : it is found, however, in some one or other of its supposed varieties, in every part of Arctic America, including the region from Hudson's Bay to far within the Arctic circle.

GENERAL REMARKS.

The American Caribou or Reindeer has by most authors been regarded as identical with the Reindeer (*Rangifer tarandus*) of Europe, Greenland, and the Asiatic polar regions. The arguments in favour of this supposition are very plausible, and the varieties which the species exhibits in America, together with the fact that the antlers of the Reindeer assume an almost infinite diversity of form, that they differ not only in different specimens, but that the horns on each side of the head of the same animal often differ from each other, afford still stronger grounds for the supposition : notwithstanding all this, supposing that they are only varieties, they have become such permanently in our continent, and require separate descriptions, and as they must be known by particular names we have supposed we might venture on designating the American Reindeer as a distinct species, admitting at the same time that the subject requires closer comparisons than we have been able to institute, and further investigations.

We believe that several naturalists have bestowed new names on the American animal, but we are not aware that any one has described it, or pointed out those peculiarities which would separate the species. Among the rest, we were informed that our esteemed friend Professor AGASSIZ had designated it as *Tarandus furcifer*, and believing that he had described it we adopted his name on our plate ; subsequently, however, we were informed that he had merely proposed for it the name of *Cervus hastatus*. He did not, however, describe it, and as the common name under which it has been known for ages past in America will be most easily understood, and can by no possibility lead to any misapprehension as regards the species, we have named it *Rangifer Caribou*, and respectfully request our subscribers to alter the name on the plate accordingly.

N°26 Plate CXXVII

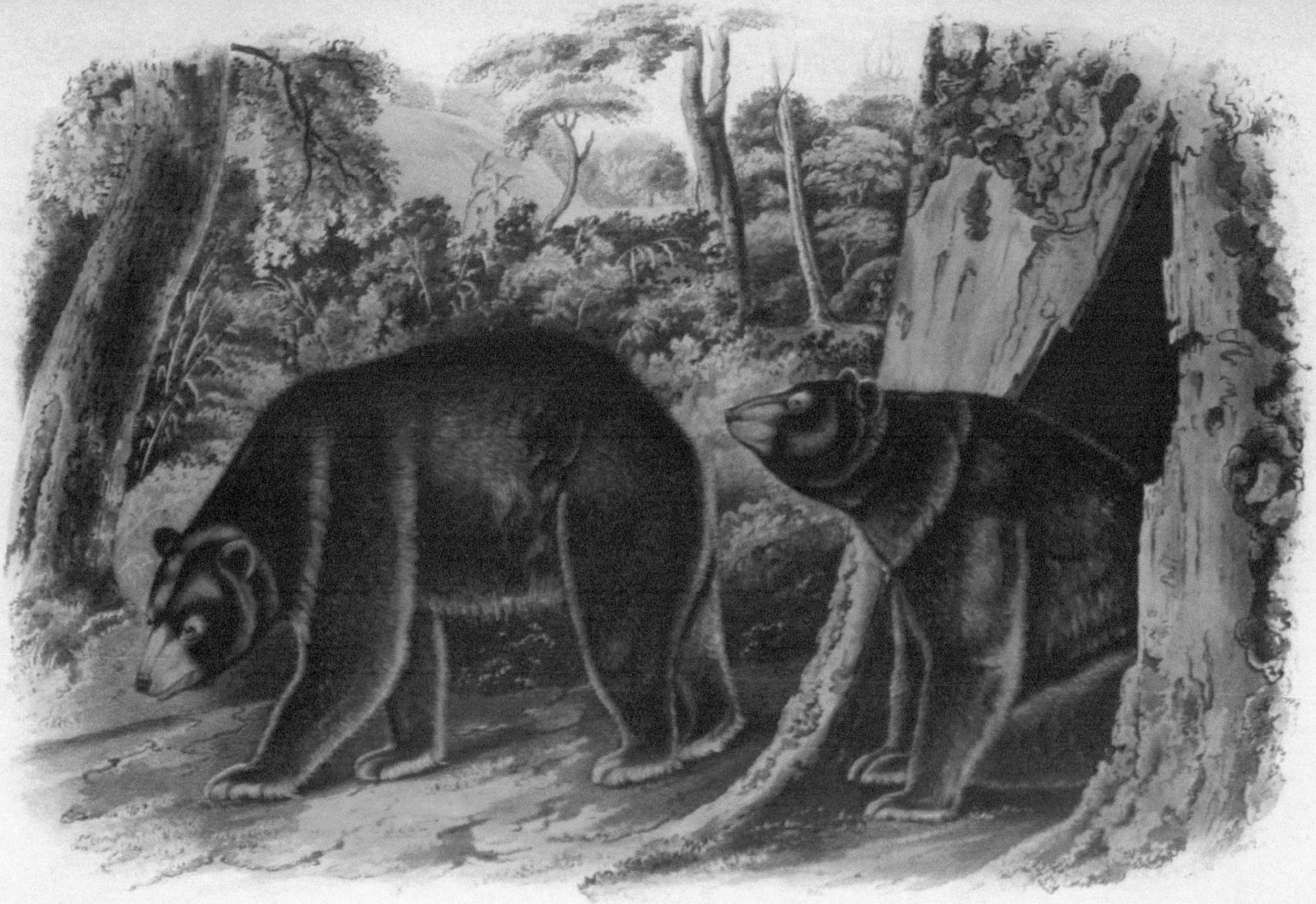

On stone by W.E. Hitchcock

Drawn from Nature by J. W. Audubon

Cinnamon Bear

Lith.d Printed & Col.d by J. T. Bowen Philad.a

URSUS AMERICANUS.—Pallas.
(Var. Cinnamomum.—Aud. and Bach.)

Cinnamon Bear.

PLATE CXXVII.—Male and Female.

U. Magnitudine formaque U. Americani ; supra saturate cinnamomeus, naso et pilis ungues vestientibus flavis.

CHARACTERS.

Form and size of the common American black bear, of which it is a permanent variety. Colour, above dark cinnamon brown ; nose and a fringe of hairs covering the claws, yellow.

SYNONYME.

Cinnamon Bear of the fur traders.

DESCRIPTION.

Form and size of the American Black Bear (*Ursus Americanus*). Hair, softer and more dense than that of the Black Bear, and under fur finer and longer.

COLOUR.

Nose, ochreous yellow ; there is an angular yellow spot above each eye ; margins of ears, and a narrow band of hairs around all the feet, concealing the claws, ochreous yellow ; there is a line of brownish-yellow from the shoulder down and along the front leg ; sides and hips, dark yellow ; a line around the cheeks from the ear downwards, and a spot and streak between the ears, a little darker yellow ; other parts of the body, cinnamon brown.

DIMENSIONS.

	Feet.	Inches.
Length from point of nose to root of tail, - - -	5	8
Height at shoulder, - - - - - - -	3	1
Length of tail, - - - - - - - -		1½

The Cinnamon Bear, like the common Black Bear, varies greatly in size. The dimensions above are unusually large.

HABITS.

LEWIS and CLARK (Expedition, vol. ii. p. 303) mention that one of their men purchased a Bear-skin " of a uniform pale reddish-brown colour, which the Indians (Chopunnish) distinguished from every variety of the Grizzly Bear : this induced those travellers to inquire more particularly into the opinions held by the Indians as to the several species of Bears, and they exhibited all the Bear-skins they had killed in that neighbourhood, which the Indians immediately classed into two species—the Grizzly Bear, including all those with the extremities of the hair of a white or frosty colour, under the name of *hohhost*, and the black skins, those which were black with a number of entire white hairs intermixed, or white with a white breast, uniform bay, brown, and light reddish-brown, were ranged under the name of *Yackkah*. These we refer to the Cinnamon and other varieties of the Black Bear. LEWIS and CLARK, however, appear not to have considered these Bear-skins as belonging to the Black Bear, owing merely to the differences in colour, for they say the common Black Bear is "indeed unknown in that country." Their account of the fur of the brown Bears above mentioned corresponds, however, with the description of the Cinnamon Bear, they remarking that the skins of the Bears in that region differ from those of the Black Bears "in having much finer, thicker, and longer hair, with a greater proportion of fur mixed with it." LEWIS and CLARK considered that the Black Bear was always black, whereas it varies very considerably : they say nothing in regard to the sizes of the various coloured Bears above alluded to.

The Cinnamon Bear has long been known to trappers and fur traders, and its skin is much more valuable than that of the Black Bear. We have seen in the warehouse of Messrs. P. CHOUTEAU, JR., and Co., in New York, some beautiful skins of this animal, and find that those gentlemen receive some every year from their posts near the Rocky Mountains. Being a permanent variety, and having longer and finer hair than the common Black Bear, we might possibly have elevated it into a distinct species but that in every other particular it closely resembles the latter animal. By the Indians (according to Sir JOHN RICHARDSON) it is considered to be an accidental variety of the Black Bear.

The Cinnamon Bear, so far as we have been able to ascertain, is never found near the sea coast, nor even west of the Ohio valley until you approach the Rocky Mountain chain, and it is apparently quite a northern animal.

Of the habits of this variety we have no accounts, but we may suppose that they do not differ in any essential particulars from those of the Black Bear, which we shall shortly describe.

Our figures were made from living specimens in the gardens of the Zoological Society of London, which manifested all the restlessness usually exhibited by this genus when in a state of captivity.

We are inclined to consider Sir JOHN RICHARDSON's "Barren-ground Bear" a variety of the common black Bear,—perhaps our present animal ; but not having seen any specimen of his *Ursus Arctos ? Americanus*, we do not feel justified in expressing more than an opinion on this subject, which indeed is founded on the description of the colour of the Barren-ground Bear as given by RICHARDSON himself (see Fauna Boreali Americana, pp. 21, 22).

GEOGRAPHICAL DISTRIBUTION.

Sparingly found in the fur countries west and north of the Missouri, extending to the barren grounds of the northwest.

GENERAL REMARKS.

We have given a figure of this permanent variety of Bear, not because we felt disposed to elevate it into a species, but because it is a variety so frequently found in the collections of skins made by our fur companies, and which is so often noticed by travellers in the northwest, that errors might be made by future naturalists were we to omit mentioning it and placing it where it should be. Whilst we are not disposed to figure an occasional variety in any species, and have throughout our work rather declined doing this, yet we conceive that figures of the permanent varieties may be useful to future observers in order to awaken inquiry and enable them to decide whether they are true species or mere varieties. We have done this in the case of some species of squirrel, the otter, and the wolves, as well as this variety of Bear. The yellow Bear of Carolina no doubt belongs to this variety, and probably the brown Barren-ground Bear of RICHARDSON may be referred to the same species, as all Bears vary very greatly in size.

GENUS CAPRA.—Linn.

Incisive $\frac{0}{8}$; *Canine* $\frac{0-0}{0-0}$; *Molar* $\frac{6-6}{6-6} = 32.$

Horns common to both sexes, or rarely wanting in the female ; in domesticated races occasionally absent in both : they are directed upwards and curved backwards, and are more or less angular. No muzzle, no lachrymal sinus, nor unguinal pores ; eyes, light coloured, pupil elongated ; tail, short, flat, and naked at base ; throat, bearded.

Mostly reside in the primitive and highest mountains of the ancient continent and America.

Habit, herbivorous ; climbing rocks and precipices ; producing two or three young at a time ; gregarious.

There are six well determined species—one inhabiting the Alps, one in Abyssinia and Upper Egypt, one in the Caucasian mountains, one in the mountains of Persia, one in the Himalaya, and one in the Rocky Mountains of North America.

The generic name Capra is derived from the Latin *capra*, a goat.

CAPRA AMERICANA.—Blainville.

Rocky Mountain Goat.

PLATE CXXVIII.—Male and Female.

C. Magnitudine ovem arietem adæquans, corpore robusto, cornibus parvis acutis lente recurvis, pilis albis, cornibus ungulisque nigris.

CHARACTERS.

Size of the domestic sheep ; form of body, robust ; horns, small and pointed, slightly curved backwards. Colour of hair, totally white.

SYNONYMES.

Antilope Americana et Rupicapra Americana. Blainville, Bulletin Socy. Phil., Ann. 1816, p. 80.
Ovis Montana. Ord, Jour. Acad. N. Sci. Phil., vol. i., part i., p. 8. Ann. 1817.

Plate CXXVIII

Drawn from Nature by J.W. Audubon.

On Stone by W.E. Hitchcock.

Lith. Printed & Col.d by J.T. Bowen, Phil.a

Rocky Mountain Goat

MAZAMA SERICEA. Raffinesque Smaltz, Am. Monthly Mag. 1817, p. 44.
ROCKY MOUNTAIN SHEEP. Jameson, Wernerian Trans., vol. iii. p. 306.   Ann. 1821.
CAPRA MONTANA. Harlan, Fauna Americana, p. 253.
  "   "   Godman, Nat. Hist., vol. ii. p. 326.
ANTELOPE LANIGERA. Smith, Linnæan Trans., vol. xiii. p. 38, t. 4.
CAPRA AMERICANA. Rich., F. B. A., p. 268, plate 22.

DESCRIPTION.

Form of the body and neck, robust, like that of the common Goat ; nose, nearly straight ; ears, pointed, lined with long hair ; the horns incline slightly backwards, tapering gradually and not suddenly, uncinated like those of the chamois, transversely wrinkled with slight rings for nearly half their length from the base, and sharp pointed ; towards the tip they are smooth and polished.   Tail short, and though clothed with long hair, almost concealed by the hairs which cover the rump ; legs, thick and short ; secondary hoofs, flat, grooved on the soles, and resembling those of the common Goat.

The coat is composed of two kinds of hair, the outer and longer considerably straighter than the wool of the sheep, but softer than that of the common Goat ; this long hair is abundant on the shoulders, back, neck, and thighs ; on the chin there is a thick tuft forming a beard like that of the latter animal ; under the long hairs of the body there is a close coat of fine white silky wool, quite equal to that of the Cashmere Goat in fineness.

COLOUR.

Horns, and hoofs, black ; the whole body, white.

DIMENSIONS.

		Feet.	Inches.
Length of head and body, - - - - -		3	4
" tail, - - - - - - -			1
" head, - - - - - - -			11
" horns, - - - - - - -			5
Diameter of horns at base, - - - - -			1

HABITS.

Standing "at gaze," on a table-rock projecting high above the valley beyond, and with a lofty ridge of stony and precipitous mountains in the background, we have placed one of our figures of the Rocky Mountain

Goat; and lying down, a little removed from the edge of the cliff, we have represented another.

In the vast ranges of wild and desolate heights, alternating with deep valleys and tremendous gorges, well named the Rocky mountains, over and through which the adventurous trapper makes his way in pursuit of the rich fur of the beaver or the hide of the bison, there are scenes which the soul must be dull indeed not to admire. In these majestic solitudes all is on a scale to awaken the sublimest emotions and fill the heart with a consciousness of the infinite Being "whose temple is all space, whose altar earth, sea, skies."

Nothing indeed can compare with the sensations induced by a view from some lofty peak of these great mountains, for there the imagination may wander unfettered, may go back without a check through ages of time to the period when an Almighty power upheaved the gigantic masses which lie on all sides far beneath and around the beholder, and find no spot upon which to arrest the eye as a place where once dwelt man! No—we only know the Indian as a wanderer, and we cannot say here stood the strong fortress, the busy city, or even the humble cot. · Nature has here been undisturbed and unsubdued, and our eyes may wander all over the scene to the most distant faint blue line on the horizon which encircles us, and forget alike the noisy clamour of toiling cities and the sweet and smiling quiet of the well cultivated fields, where man has made a "home" and dwelleth in peace. But in these regions we may find the savage grizzly bear, the huge bison, the elegant and fleet antelope, the large-horned sheep of the mountains, and the agile fearless climber of the steeps—the Rocky Mountain Goat.

This snow-white and beautiful animal appears to have been first described, from skins shown to LEWIS and CLARK, as "the Sheep," in their general description of the beasts, birds, and plants found by the party in their expedition. They say, "The Sheep is found in many places, but mostly in the timbered parts of the Rocky Mountains. They live in greater numbers on the chain of mountains forming the commencement of the woody country on the coast, and passing the Columbia between the falls and the rapids. We have only seen the skins of these animals, which the natives dress with the wool, and the blankets which they manufacture from the wool. The animal from this evidence appears to be of the size of our common sheep, of a white colour. The wool is fine on many parts of the body, but in length not equal to that of our domestic sheep. On the back, and particularly on the top of the head, this is intermixed with a considerable portion of long straight hairs. From the Indian account these animals have erect pointed horns."

The Rocky Mountain Goat wanders over the most precipitous rocks, and springs with great activity from crag to crag, feeding on the plants, grasses, and mosses of the mountain sides, and seldom or never descends to the luxuriant valleys, as the Big-Horn does. This Goat indeed resembles the wild Goat of Europe, or the chamois, in its habits, and is very difficult to procure. Now and then the hunter may observe one browsing on the extreme verge of some perpendicular rock almost directly above him, far beyond gun-shot, and entirely out of harm's way. At another time, after fatiguing and hazardous efforts, the hungry marksman may reach a spot from whence his rifle will send a ball into the unsuspecting Goat; then slowly he rises from his hands and knees, on which he has been creeping, and the muzzle of his heavy gun is "rested" on a loose stone, behind which he has kept his movements from being observed, and now he pulls the fatal trigger with deadly aim. The loud sharp crack of the rifle has hardly rung back in his ear from the surrounding cliffs when he sees the Goat in its expiring struggles reach the verge of the dizzy height: a moment of suspense and it rolls over, and swiftly falls, striking perchance here and there a projecting point, and with the clatter of thousands of small stones set in motion by its rapid passage down the steep slopes which incline outward near the base of the cliff, disappears, enveloped in a cloud of dust in the deep ravine beneath; where a day's journey would hardly bring an active man to it, for far around must he go to accomplish a safe descent, and toilsome and dangerous must be his progress up the gorge within whose dark recesses his game is likely to become the food of the ever prowling wolf or the solitary raven. Indeed cases have been mentioned to us in which these Goats, when shot, fell on to a jutting ledge, and there lay fifty or a hundred feet below the hunter, in full view, but inaccessible from any point whatever.

Notwithstanding these difficulties, as portions of the mountains are not so precipitous, the Rocky Mountain Goat is shot and procured tolerably easily, it is said, by some of the Indian tribes, who make various articles of clothing out of its skin, and use its soft woolly hair for their rude fabrics.

According to Sir JOHN RICHARDSON, this animal has been known to the members of the Northwest and Hudson's Bay Companies from the first establishment of their trading posts on the banks of the Columbia River and in New Caledonia, and they have sent several specimens to Europe. The wool being examined by a competent judge, under the instructions of the Wernerian Society of Edinburgh, was reported to be of great fineness and fully an inch and a half long. "It is unlike the fleece of the common sheep, which contains a variety of different kinds of wool suitable to the fabrication of articles very dissimilar in their nature, and requires much care to distribute them in their proper order. The fleece under consideration is wholly fine. That on the fore part of the skin has all the apparent qualities of wool. On the back part it very much resembles cotton. The whole fleece is much mixed with hairs, and on those parts where the hairs are long and pendant, there is almost no wool."

" Mr. DRUMMOND saw no Goats on the eastern declivity of the mountains, near the sources of the Elk river, where the sheep are numerous, but he learned from the Indians that they frequent the steepest precipices, and are much more difficult to procure than the sheep. Their manners are said to greatly resemble those of the domestic Goat. The exact limits of the range of this animal have not been ascertained, but it probably extends from the fortieth to the sixty-fourth or sixty-fifth degree of latitude. It is common on the elevated part of the Rocky Mountain range that gives origin to four great tributaries to as many different seas, viz. the Mackenzie, the Columbia, the Nelson, and the Missouri rivers."—*F. B. A.*, p. 269.

The flesh of this species is hard and dry, and is not so much relished as that of the Big-Horn, the Elk, &c., by the hunters or travellers who have journeyed towards the Pacific across the wild ranges of mountains inhabited by these animals.

GEOGRAPHICAL DISTRIBUTION.

The Rocky Mountain Goat inhabits the most elevated portions of the mountains from which it derives its name, where it dwells between the fortieth and sixtieth or sixty-fourth degree of north latitude. It is also found on the head waters of the Mackenzie, Columbia, and Missouri rivers. Mr. MACKENZIE informs us that the country near the sources of the Muddy river (Maria's river of LEWIS and CLARK), Saskatchewan, and Athabasca, is inhabited by these animals, but they are said to be scarcer on the eastern slopes of the Rocky Mountains than on the western.

GENERAL REMARKS.

It is believed by some naturalists that Fathers PICCOLO and DE SALVATIERRA discovered this animal on the higher mountains of California. VANCOUVER brought home a mutilated skin which he obtained on the northwest coast of America. LEWIS and CLARK (as we have already mentioned) obtained skins in 1804.

In 1816 M. DE BLAINVILLE published the first scientific account of it. Mr. ORD in 1817 described one of the skins brought home by LEWIS and

CLARK, and Major CHARLES HAMILTON SMITH described a specimen in 1821, in the Linnæan Transactions for that year.

The resemblance of the animal to some of the antelopes, the chamois, the Goat, and the sheep, caused it to be placed by these authors under several genera. DE BLAINVILLE first made it an *antelope*, then named it *Rupi-capra*—a subgenus of antelope to which the chamois belongs. ORD arranged it in the genus *Ovis*. SMITH called it *Antilope lanigera*. Besides these, RAFFINESQUE named it *Mazama sericea*. Dr. HARLAN and RICHARDSON were each correct, as we think, in placing it in the genus *Capra* (Goat). As in the Goat, the facial line in this species is nearly straight, while in the sheep and antelopes it is more or less arched. The sheep and the antelope are beardless, and the Goat is characterized by its beard, a conspicuous ornament in the present animal, which is moreover, in the form of its nose, the strength and proportion of the limbs, and the peculiarities of the hoofs, allied closer to the Goats than to any other neighbouring genus.

ARVICOLA BOREALIS.—Rich.

Northern Meadow-Mouse.

PLATE CXXIX.—Male and Females.

A. ungue pollicari robusto præditus, auriculis vellere absconditis, cauda capitis fere longitudine, vellere longissima molli, dorso castaneo nigro mixto, ventre cano.

CHARACTERS.

Thumb nail, strong ; ears, concealed in the fur ; tail, about as long as the head ; fur, very long and fine ; on the back, chesnut colour mixed with black ; on the belly, gray.

SYNONYMES.

Mouse No. 15. Forster, Philos. Trans., vol. lxii. p. 380.
Arvicola Borealis. Rich., Zool. Jour., No. 12, April, 1828, p. 517.
   "   "   "   Northern Meadow-Mouse. F. B. A., p. 127.
Arvinnak. Dog-Rib Indians.

DESCRIPTION.

This species is a little less than Wilson's Meadow-Mouse (*A. Pennsylvanica*). It has the form and dentition of the other species of Arvicolæ. Head, rather large ; forehead, convex ; nose, short, and a little pointed ; eyes, small ; ears, low, rounded, and concealed by the surrounding fur ; limbs, rather robust, clothed with short hairs, mixed on the toes and hind parts of the fore feet with longer hairs. Hind toes, more slender, and scarcely longer than the fore ones ; fore claws, small, much compressed, arched, and acute, with a narrow elliptical excavation underneath ; the hairs of the toes reach to the points of the nails, but cover them rather sparingly ; the claws of the hind feet resemble those of the fore feet, but are not so strong ; the thumb of the fore feet consists of a small squarish nail slightly convex on both sides, and having an obtuse point projecting from the middle of its extremity ; the tail is round, well clothed with short stiff hairs running to a point, which do not permit the scales to be visible. There are considerable variations in the length of the tail, it being in one specimen a third longer than in others. The fur on the body is long in proportion to the size of the animal.

PLATE CXXIX

Northern Meadow Mouse

COLOUR.

Hair on the upper parts blackish-gray from the roots to the tips, some of which are yellowish or chesnut brown, and some black ; the black tipped hairs are the longest, and are equally distributed amongst the others, giving the body a dark reddish-brown colour. There is a rufous patch under the ears. On the under part, and on the chin and lips, the colour is lead-gray, and the hairs are shorter than on the back and sides ; tail, brown above and grayish beneath ; hairs on the feet, ochreous yellow ; claws, white.

DIMENSIONS.

	Inches.	Lines.
Length of head and body, - - - - - - 4		6
" tail, - - - - - - - 1		
" head, - - - - - - - 1		3
Height of ear, - - - - - - -		4
Breadth of ear, - - - - - - -		3
Length of fore feet to end of middle claw, - - -		4½
Hind feet, including heel and claw, - - - -		7½
Fur on the back, - - - - - - -		10

HABITS.

We have little to say in regard to the present species. RICHARDSON states that its habits are very similar to those of *A. xanthognatha*, and in our article on that species we have given an account of the general habits of the Arvicolæ (at p. 18 of the present volume), to which we refer our readers.

The northern Arvicolæ do not appear to become dormant from the effect of cold, but during the long Arctic winter dig galleries under the deep snows, in which they are enabled to search for seeds, grasses, or roots suited to their wants. We have ascertained by an examination of the bodies of several, more southern species of Arvicolæ, possessing similar habits, that so far from suffering in winter and becoming lean, they are usually in good case, and sometimes quite fat, during that season.

The length of the fur on the back of the present species (ten twelfths of an inch) is somewhat remarkable for so small an animal.

GEOGRAPHICAL DISTRIBUTION.

This species was found in numbers at Great Bear Lake, living in the vicinity of *Arvicola xanthognatha*. We have not been able to ascertain the extent of its range towards the south or west. We did not discover this Meadow-Mouse or hear of it on our expedition to the Yellow Stone and Upper Missouri rivers, nor has it been found, so far as we know, anywhere west of the Rocky Mountains.

GENERAL REMARKS.

"The form of the thumb-nail allies this animal very closely to the Norway lemming, and to one or two species of American lemming, but its claws are smaller and more compressed, and apparently not so well calculated for scraping earth as the broader claws of the lemmings."—*Fauna Boreali Americana*, p. 127.

Thus far we agree with Dr. RICHARDSON ; he, however, thinks that this species may be considered an intermediate link between the lemmings and the Meadow-Mice, and may without impropriety be ranked either as a true Meadow-Mouse or as a lemming.

After a careful examination of the original specimens, some years ago, we set it down as a true *Arvicola*, possessing more of the characteristics of that genus than of the genus *Georychus*.

Plate LXXVI

On Stone by Wm. E. Hitchcock.

Pouched Jerboa Mouse.

Lith Printed & Cold by J.T. Bowen, Phil.

GENUS DIPODOMYS.—Gray.

DENTAL FORMULA.

Incisive $\frac{2}{2}$; *Canine* $\frac{0-0}{0-0}$; *Molar* $\frac{4-4}{4-4} = 20$.

The incisors are of moderate length, rather weak, narrow, compressed, and curved inwards. In the upper jaw the first three molars are largest, the fourth a little smaller; in the lower jaw the molars are alike. The molars have rounded cutting edges.

Nose and head, of moderate size; sacs or pouches opening on the cheeks back of the mouth; fore feet, rather short, furnished with four toes and the rudiment of a thumb, covered by a blunt nail; hind legs very long, terminated by four toes on each foot; toes, each with a distinct metatarsus; tail, very long; *mammæ*, four—two abdominal and two pectoral.

Habits, semi-nocturnal; food, seeds, roots, and grasses.

There is only one species belonging to this genus known. The generic name is derived from δίπους, *dipous*, two footed, and μυς, *mus*, a mouse.

DIPODOMYS PHILLIPPSII.—Gray.

Pouched Jerboa Mouse.

PLATE CXXX.—Males.

D. Magnitudine prope Tamiæ Lysteri et formâ Dipodum; caudâ corpore et capite conjunctum multo longiore; sacculis buccalibus externis apertis; colore, supra fulvo, infra albo.

CHARACTERS.

Nearly the size of the common ground squirrel (Tamias Lysteri); *shaped like the jerboas; tail, much longer than the body; cheek pouches, opening externally; colour, light brown above, white beneath.*

SYNONYME.

Dipodomys Phillippsii. Gray, Ann. and Mag. Nat. Hist., vol. vii. p. 521. 1840.

DESCRIPTION.

Body, rather stout; head, of moderate size; nose, moderate, although the skull exhibits the proboscis extended five or six lines beyond the insertion of the incisors.

The whiskers (which proceed from the nose immediately above the upper edge of the orifices of the pouches) are numerous, rigid, and longer than the head; ears, of moderate size, ovate, and very thinly clothed with short hairs; the feet are thickly clothed with short hairs to the nails, which are free; short hairs also prevail on the soles and between the toes; fore feet, rather stout, but short; they have each four toes and the rudiment of a thumb, the latter covered by a conspicuous nail; nails, short, slender, and curved; second toe from the thumb longest, first and third nearly of equal length, and fourth shortest.

Hind legs, very long; the hind feet have each four toes, the two middle ones nearly of equal length, the first a little shorter, and the fourth, placed behind like a thumb, much the shortest; nails, nearly straight, sharp pointed, and grooved on the under surface; tail, rather stout—in the dried specimen it is round at base and much compressed, showing that its greatest diameter is vertical; it is thickly clothed with short hairs for two thirds of its extent, when the hairs gradually increase in length till they approach the extremity, at which they are so long as to present the appearance of a tuft-like brush. The fur is very soft and silky, like that of the flying-squirrel; the hairs of the tail are coarser. There are two abdominal and two pectoral mammæ.

In the upper jaw the incisors are rather small and weak; all the molars have simple crowns, which are more elevated on the interior than on the exterior edges; the anterior molar is nearly round, and almost of the same size as the two next molars, which are somewhat oval and are placed with their longest diameter transversely to the jaw; the fourth molar is the smallest and is nearly round.

In the lower jaw the three anterior molars are nearly of equal size, and are almost alike in shape; the fourth corresponding with the last molar on the upper jaw; there is a little depression in the centre of the crowns of the molars, and a slight ridge around the outer edges.

COLOUR.

Head, ears, back, and a stripe on the thigh from the root of the tail, light brown, the hairs on the back being plumbeous at the roots, then yellow slightly tipped with black. Whiskers, black, with a few white oristly hairs interspersed ; upper and lower surfaces of tail, and a line on the under side of the tarsus, dark brown ; sides, and tip of the tail, white ; cheeks, white ; there is a white stripe on the hips : the legs and under surface are white, as also a stripe from the shoulder to the ear. This white colour likewise extends high up on the flanks, where it gradually mingles with the brown of the back ; nails, brownish.

DIMENSIONS.

Male.—Specimen in the British Museum.

	Inches.	Lines.
Length of head and body, - - - - -	5	
" tail, - - - - - - -	6	6
" hind feet, - - - - - -	1	6

Female.—Procured by J. W. Audubon in California.

	Inches.	Lines
Point of nose to root of tail, - - - -	4	6
Tail, including hair, - - - - - -	7	
Tarsus to end of longest nail, - - - -	1	6
Ear, inside, from auditory opening, - - -		7
Longest hair of whiskers, - - - - -	2	4

HABITS.

The pretty colours and the liveliness of this little kangaroo-like animal, together with its fine eyes and its simplicity in venturing near man, of whom it does not seem afraid, would no doubt make it a favourite pet in confinement. It is able to exist in very arid and almost barren situations, where there is scarcely a blade of anything green except the gigantic and fantastic cacti that grow in Sonora and various other parts of Western Mexico and California. As John W. Audubon and his party travelled through these countries the *Dipodomys Phillippsii* was sometimes almost trampled on by the mules, and was so tame that they could have caught the animal by the hand without difficulty.

This species hop about, kangaroo fashion, and jump pretty far at a leap. When the men encamped towards evening, they sometimes came smelling and moving about the legs of the mules, as if old friends. One was observed by J. W. Audubon just before sunset ; its beautiful large eyes seemed as if they might be dimmed by the bright rays which fell upon them as it emerged from a hole under a large boulder, but it frisked gaily about, and several times approached him so nearly, as he sat on a stone, that he could have seized it with his hands without any trouble, and without rising from his hard seat.

After a while, as the party had to take up the line of march again, he with some difficulty frightened it, when with a bound or two it reached its hole and disappeared underneath the large stone, but almost immediately came out again ; and so great was its curiosity that as the party left the spot it seemed half inclined to follow them.

These animals appear to prefer the sides of stony hills which afford them secure places to hide in, and they can easily convey their food in their cheek-pouches to their nests.

The young when half grown exhibit the markings of the adults to a great extent. This species is crepuscular if not nocturnal, and was generally seen towards dusk, and occasionally in such barren deserts that it was difficult to imagine what it could get to feed on. A dead one was picked up one day while the party were traversing a portion of the great Colorado desert, where nothing could grow but clumps of cacti of different species, and not a drop of water could be found. The only living creatures appeared to be lizards of several kinds, and one or two snakes : the party felt surprised as they toiled on over the sun-baked clay, and still harder gravel, to find the little animal in such a locality.

GEOGRAPHICAL DISTRIBUTION.

Dr. J. L. Le Conte found this species on the river Gila, and farther south, where he procured several specimens.

J. W. Audubon saw the *Dipodomys Phillippsii* in crossing the Cordilleras, in Sonora on the Gila, in the Tulare valley, and in various other parts of California. Its southern limits are undetermined, but it seems not to exist north of California.

GENERAL REMARKS.

Mr. Gray described this species, in the Annals and Magazine of Natural History, vol. vii. p. 521 ; he considered it the American representative of the African Jerboas, although, as he remarks, it differs from them in being provided with cheek pouches opening externally.

Our drawing was made from a beautiful specimen in the British Museum, which was the first one brought under the notice of naturalists, and the original of Mr. Gray's description of this singular animal ; it was procured near Real del Monte, in Mexico.

URSUS FEROX.—Lewis and Clark.

Grizzly Bear.

PLATE CXXXI.—Males.

M. Magnitudine U. Americanum longe superans, plantis et unguibus longioribus, auriculis brevioribus quam in isto ; pilis saturate fuscis, apice griseis.

CHARACTERS.

Larger than the American Black Bear ; soles of feet, and claws, longer, and ears shorter than in the Black Bear. Colour of the hair, dark brown, with paler tips.

SYNONYMES.

Grizzle Bear. Umfreville, Hudson's Bay, p. 168. Ann. 1790.

Grisly Bear. Mackenzie's Voyage, p. 160. Ann. 1801.

White, or Brown-grey Bear. Gass' Journal of Lewis and Clark's Expedition, pp. 45, 116, 346. Ann. 1808.

Grizzly, Brown, White, and Variegated Bear—Ursus Ferox. Lewis and Clark, Expedition, vol. i. pp. 284, 293, 343, 375; vol. iii. pp. 25, 268. Ann. 1814.

Ursus Ferox. De Witt Clinton, Trans. Philos. and Lit. Society New York, vol. i p. 56. Ann. 1815.

Grizzly Bear. Warden's United States, vol. i. p. 197. Ann. 1819.

Grey Bear. Harmon's Journal, p. 417. Ann. 1820.

Ursus Cinereus. Desm. Mamm. No. 253. Ann. 1820.

" Horribilis. Ord, Guthrie's Geography, vol. ii. p. 299.

" " Say, Long's Expedition, vol. ii. p. 244, note 84. Ann. 1822.

" Candescens. Hamilton Smith, Griffith An. Kingdom, vol. ii. p. 299 ; vol. v. No. 320. Ann. 1826.

" Cinereus. Harlan, Fauna, p. 48.

Grizzly Bear. Godman's Nat. Hist., vol. i. p. 131.

Ursus Ferox. Rich., Fauna Boreali Americana, p. 24, plate 1.

DESCRIPTION.

The Grizzly Bear in form resembles the Norwegian variety of *Ursus Arctos*, the Brown Bear of Europe ; the facial line is rectilinear or slightly arched ; head, short and round ; nose, bare ; ears, rather small, and more hairy than those of the Black Bear; legs, stout; body, large, but less fat and heavy in proportion, than that of the Black Bear.

Tail, short; paws and nails, very long, the latter extending from three to five inches beyond the hair on the toes; they are compressed and channelled. Hair, long and abundant, particularly about the head and neck, the longest hairs being in summer about three inches, and in winter five or six inches long. The jaws are strong, and the teeth very large.

The fore feet somewhat resemble the human hand, and are soft to the touch; they have larger claws than the hind feet. The animal treads on the whole palm and entire heel.

COLOUR.

The Grizzly Bear varies greatly in colour, so much so, indeed, that it is difficult to find two specimens alike: the young are in general blacker than the old ones. The hair however is commonly dark brown at the roots and for about three fourths of its length, then gradually fades into reddish-brown, and is broadly tipped with white intermixed with irregular patches of black or dull-brown, thus presenting a hoary or grizzly appearance on the surface, from which the vulgar specific name is derived.

A specimen procured by us presents the following colouring: Nose, to near the eyes, light brown; legs, forehead, and ears, black. An irregularly mixed dark grayish-brown prevails on the body, except on the neck, shoulders, upper portion of fore-legs, and sides adjoining the shoulders, which parts are barred or marked with light yellowish-gray, and the hairs in places tipped with yellowish or dingy white. Iris, dark brown.

DIMENSIONS.

Male, killed by J. J. AUDUBON and party on the Missouri river, in 1843—not full grown.

	Feet.	Inches.
From point of nose to root of tail, - - -	5	6
Tail (vertebræ), - - - - - -		3
" (including hair), - - - - -		4
From point of nose to ear, - - - - -	1	4
Width of ear, - - - - - -		3½
Length of eye, - - - - - -		1
Height at shoulder, - - - - -	3	5
" rump, - - - - -	4	7
Length of palm of fore foot, - - - -		8
Breadth of do., - - - - - -		6

	Feet.	Inches.
Length of sole of hind foot, - - - - -		9½
Breadth of do., - - - - - - -		5½
Girth around the body, behind the shoulders, -	4	1
Width between the ears on the skull, - - -		7½

HABITS.

We have passed many hours of excitement, and some, perchance, of danger, in the wilder portions of our country ; and at times memory recals adventures we can now hardly attempt to describe ; nor can we ever again feel the enthusiasm such scenes produced in us. Our readers must therefore imagine, the startling sensations experienced on a sudden and quite unexpected face-to-face meeting with the savage Grizzly Bear—the huge shaggy monster disputing possession of the wilderness against all comers, and threatening immediate attack !

Whilst in a neighbourhood where the Grizzly Bear may possibly be hidden, the excited nerves will cause the heart's pulsations to quicken if but a startled ground-squirrel run past ; the sharp click of the lock is heard, and the rifle hastily thrown to the shoulder, before a second of time has assured the hunter of the trifling cause of his emotion.

But although dreaded alike by white hunter and by red man, this animal is fortunately not very abundant to the eastward of the Rocky Mountains, and the chance of encountering him does not often occur. We saw only a few of these formidable beasts during our expedition up the Missouri river and in the country over which we hunted during our last journey to the west.

The Indians, as is well known, consider the slaughter of a Grizzly Bear a feat second only to scalping an enemy, and necklaces made of the claws of this beast are worn as trophies by even the bravest among them.

On the 22d of August, 1843, we killed one of these Bears, and as our journals are before us, and thinking it may be of interest, we will extract the account of the day's proceedings, although part of it has no connection with our present subject. We were descending the Upper Missouri river.

" The weather being fine we left our camp of the previous night early, but had made only about twelve miles when the wind arose and prevented our men from making any headway with the oars ; we therefore landed under a high bank amongst a number of fallen trees and some drifted timber. All hands went in search of elks. Mr. CULBERTSON killed a deer, and with the help of Mr. SQUIRES brought the meat to the boat. We saw nothing during a long walk we took, but hearing three or four gunshots which we thought were fired by some of our party, we hastened in the direction from whence the reports came, running and hallooing, but could find no one. We then made the best of our way back to the boat and despatched three men, who discovered that the firing had been at an elk, which was however not obtained. Mr. BELL killed a female elk and brought a portion of its flesh to the boat. After resting ourselves a while and eating dinner, Mr. CULBERTSON, SQUIRES, and ourselves walked to the banks of the Little Missouri, distant about one mile, where we saw a buffalo bull drinking at the edge of a sand-bar. We shot him, and fording the stream, which was quite shallow, took away the 'nerf;' the animal was quite dead. We saw many ducks in this river. In the course of the afternoon we started in our boat, and rowed about half a mile below the Little Missouri. Mr. CULBERTSON and ourselves walked to the body of the bull again and knocked off his horns, after which Mr. CULBERTSON endeavoured to penetrate a large thicket in hopes of starting a Grizzly Bear, but found it so entangled with briars and vines that he was obliged to desist, and returned very soon Mr. HARRIS, who had gone in the same direction and for the same purpose, did not return with him. As we were approaching the boat we met Mr. SPRAGUE, who informed us that he thought he had seen a Grizzly Bear walking along the upper bank of the river, and we went towards the spot as fast as possible. Meantime the Bear had gone down to the water, and was clumsily and slowly proceeding on its way. It was only a few paces from and below us, and was seen by our whole party at the same instant. We all fired, and the animal dropped dead without even the power of uttering a groan. Mr. CULBERTSON put a rifle ball through its neck, BELL placed two large balls in its side, and our bullet entered its belly. After shooting the Bear we proceeded to a village of 'prairie dogs' (*Spermophilus Ludovicianus*), and set traps in hopes of catching some of them. We were inclined to think they had all left, but Mr. BELL seeing two, shot them. There were thousands of their burrows in sight. Our 'patroon,' assisted by one of the men, skinned the Bear, which weighed, as we thought, about four hundred pounds. It appeared to be between four and five years old, and was a male. Its lard was rendered, and filled sundry bottles with 'real Bear's grease,' whilst we had the skin preserved by our accomplished taxidermist, Mr. BELL."

The following afternoon, as we were descending the stream, we saw another Grizzly Bear, somewhat smaller than the one mentioned above. It was swimming towards the carcase of a dead buffalo lodged in the prongs of a "sawyer" or "snag," but on seeing us it raised on its hind feet until quite erect, uttered a loud grunt or snort, made a leap from the water, gained the upper bank of the river, and disappeared in an

Plate LXXXI

Drawn from Nature by J.W. Audubon.

On stone by W.E. Hitchcock.

Lith Printed & Col.d by J.T. Bowen, Philad.

Grizzly Bear.

instant amid the tangled briars and bushes thereabouts. Many wolves of different colours—black, white, red, or brindle—were also intent on going to the buffalo to gorge themselves on the carrion, but took fright at our approach, and we saw them sneaking away with their tails pretty close to their hind-legs."

The Grizzly Bear generally inhabits the swampy, well covered portions of the districts where it is found, keeping a good deal among the trees and bushes, and in these retreats it has its "beds" or lairs. Some of these we passed by, and our sensations were the reverse of pleasant whilst in such thick, tangled, and dangerous neighbourhoods ; the Bear in his concealment having decidedly the advantage in case one should come upon him una-wares. These animals ramble abroad both by day and night. In many places we found their great tracks along the banks of the rivers where they had been prowling in search of food. There are seasons during the latter part of summer, when the wild fruits that are eagerly sought after by the Bears are very abundant. These beasts then feed upon them, tear-ing down the branches as far as they can reach whilst standing in an upright posture. They in this manner get at wild plums, service berries, buffalo berries, and the seeds of a species of *cornus* or dog-wood which grows in the alluvial bottoms of the northwest. The Grizzly Bear is also in the habit of scratching the gravelly earth on the sides of hills where the vegetable called "pomme blanche" is known to grow, but the favourite food of these animals is the more savoury flesh of such beasts as are less powerful, fleet, or cunning than themselves. They have been known to seize a wounded buffalo, kill it, and partially bury it in the earth for future use, after having gorged themselves on the best parts of its flesh and lapped up the warm blood.

We have heard many adventures related, which occurred to hunters either when surprised by these Bears, or when approaching them with the intention of shooting them. A few of these accounts, which we believe are true, we will introduce : During a voyage (on board one of the steamers belonging to the American Fur Company) up the Missouri river, a large she-Bear with two young was observed from the deck, and several gentlemen proposed to go ashore, kill the dam, and secure her cubs. A small boat was lowered for their accommodation, and with guns and ammunition they pushed off to the bank and landed in the mud. The old Bear had observed them and removed her position to some distance, where she stood near the bank, which was there several feet above the bed of the river. One of the hunters having neared the animal, fired at her, inflicting a severe wound. Enraged with pain the Bear rushed with open jaws towards the sportsmen at a rapid rate, and with looks that assured them she was in a desperate fury. There was but a moment's time; the party, too much frightened to stand the charge, "ingloriously turned and fled," without even pulling another trigger, and darting to the margin of the river jumped into the stream, losing their guns, and floundering and bobbing under, while their hats floated away with the muddy current. After swimming a while they were picked up by the steamer, as terrified as if the Bear was even then among them, though the animal on seeing them all afloat had made off, followed by her young.

The following was related to us by one of the " engagés" at Fort Union : A fellow having killed an Indian woman, was forced to run away, and fearing he would be captured, started so suddenly that he took neither gun nor other weapon with him ; he made his way to the Crow Indians, some three hundred miles up the Yellow Stone river, where he arrived in a miserable plight, having suffered from hunger and exposure. He escaped the men who were first sent after him, by keeping in ravines and hiding closely ; but others were despatched, who finally caught him. He said that one day he saw a dead buffalo lying near the river bank, and going towards it to get some of the meat, to his utter astonishment and horror a young Grizzly Bear which was feeding on the carcass, raised up from behind it and so suddenly attacked him that his face and hands were lacerated by its claws before he had time to think of defending himself. Not daunted, however, he gave the cub a tremendous jerk, which threw it down, and took to his heels, leaving the young savage in possession of the prize.

The audacity of these Bears in approaching the neighbourhood of Fort Union at times was remarkable. The waiter, " Jean Battiste," who had been in the employ of the company for upwards of twenty years, told us that while one day picking peas in the garden, as he advanced towards the end of one of the rows, he saw a large Grizzly Bear gathering that excellent vegetable also. At this unexpected and startling discovery, he dropped his bucket, peas and all, and fled at his fastest pace to the Fort. Immediately the hunters turned out on their best horses, and by riding in a circle, formed a line which enabled them to approach the Bear on all sides. They found the animal greedily feasting on the peas, and shot him without his apparently caring for their approach. We need hardly say the bucket was empty.

In GODMAN's Natural History there are several anecdotes connected with the Grizzly Bear. The first is as follows : A Mr. JOHN DOUGHERTY, a very experienced and respectable hunter belonging to Major LONG's expedition, relates that once, while hunting with another person on one of the upper tributaries of the Missouri, he heard the report of his companion's rifle, and when he looked round, beheld him at a short distance endeavouring to escape from one of these beasts, which he had wounded as it was coming towards him. DOUGHERTY, forgetful of every thing but the preservation of his friend, hastened to call off the attention of the Bear, and arrived in rifle-shot distance just in time to effect his generous purpose. He discharged his ball at the animal, and was obliged in his turn to fly ; his friend, relieved from immediate danger, prepared for another attack by charging his rifle, with which he again wounded the Bear, and saved Mr. DOUGHERTY from peril. Neither received any injury from this encounter, in which the Bear was at length killed.

On another occasion, several hunters were chased by a Grizzly Bear, which rapidly gained upon them. A boy of the party, who could not run so fast as his companions, perceiving the Bear very near him, fell with his face towards the ground. The animal reared up on his hind feet, stood for a moment, and then bounded over him, impatient to catch the more distant fugitives.

Mr. DOUGHERTY, the hunter before mentioned, relates the following instance of the great muscular strength of the Grizzly Bear : Having killed a bison, and left the carcass for the purpose of procuring assistance to skin and cut it up, he was very much surprised on his return to find that it had been dragged off whole, to a considerable distance by a Grizzly Bear, and was then placed in a pit which the animal had dug with his claws for its reception.

The following is taken from Sir JOHN RICHARDSON's Fauna Boreali Americana : "A party of voyagers, who had been employed all day in tracking a canoe up the Saskatchewan, had seated themselves in the bright light by a fire, and were busy in preparing their supper, when a large Grizzly Bear sprung over their canoe, that was placed behind them, and seizing one of the party by the shoulder, carried him off. The rest fled in terror, with the exception of a Metis, named BOURAPO, who, grasping his gun, followed the Bear as it was retreating leisurely with its prey. He called to his unfortunate comrade that he was afraid of hitting him if he fired at the Bear, but the latter entreated him to fire immediately, without hesitation, as the Bear was squeezing him to death. On this he took a deliberate aim and discharged the contents of his piece into the body of the Bear, which instantly dropped its prey to pursue BOURAPO. He escaped with difficulty, and the Bear ultimately retired to a thicket, where it was supposed to have died ; but the curiosity of the party not being a match for their fears, the fact of its decease was not ascertained. The man who was rescued had his arm fractured, and was otherwise severely bitten by the Bear, but finally recovered. I have seen BOURAPO, and can add that the account which he gives is fully credited by the traders resident in that part of the country, who are best qualified to judge of its truth from the knowledge of the parties. I have been told that there is a man now living in the neighbourhood of Edmonton-house who was attacked by a Grizzly Bear, which sprang out of a thicket, and with one stroke of its paw completely scalped him, laying bare the skull and bringing the skin of the forehead down over the eyes. Assistance coming up, the Bear made off without doing him further injury, but the scalp not being replaced, the poor man has lost his sight, although he thinks that his eyes are uninjured."

Mr. DRUMMOND, in his excursions over the Rocky Mountains, had frequent opportunities of observing the manners of the Grizzly Bear, and it often happened that in turning the point of a rock or sharp angle of a valley, he came suddenly upon one or more of them. On such occasions they reared on their hind legs and made a loud noise like a person breathing quick, but much harsher. He kept his ground without attempting to molest them, and they, on their part, after attentively regarding him for some time, generally wheeled round and galloped off, though, from their disposition, there is little doubt but he would have been torn in pieces had he lost his presence of mind and attempted to fly. When he discovered them from a distance, he generally frightened them away by beating on a large tin box, in which he carried his specimens of plants. He never saw more than four together, and two of these he supposes to have been cubs ; he more often met them singly or in pairs. He was only once attacked, and then by a female, for the purpose of allowing her cubs time to escape. His gun on this occasion missed fire, but he kept her at bay with the stock of it, until some gentlemen of the Hudson's Bay Company, with whom he was travelling at the time, came up and drove her off. In the latter end of June, 1826, he observed a male caressing a female, and soon afterwards they both came towards him, but whether accidentally, or for the purpose of attacking him, he was uncertain. He ascended a tree, and as the female drew near, fired at and mortally wounded her. She uttered a few loud screams, which threw the male into a furious rage, and he reared up against the trunk of the tree in which Mr. DRUMMOND was seated, but never attempted to ascend it. The female, in the meantime, retired to a short distance, lay down, and as the male was proceeding to join her, Mr. DRUMMOND shot him also.

The young Grizzly Bears and gravid females hibernate, but the older males often come abroad in the winter in quest of food. MACKENZIE mentions the den or winter retreat of a Grizzly Bear, which was ten feet wide, five feet high, and six feet long.

This species varies very much in colour ; we have skins in our possession collected on the Upper Missouri, some of which are nearly white, whilst others are as nearly of a rufous tint. The one that was killed by our party (of which we have also the skin) was a dark brown one.

The following is from notes of J. W. AUDUBON, made in California in 1849 and 1850 : "High up on the waters of the San Joaquin, in California, many of these animals have been killed by the miners now overrunning all the country west of the Sierra Nevada. Greatly as the Grizzly Bear is dreaded, it is hunted with all the more enthusiasm by these fearless pioneers in the romantic hills, valleys, and wild mountains of the land of gold, as its flesh is highly prized by men who have been living for months on salt pork or dry and tasteless deer-meat. I have seen two dollars a pound paid for the leaf-fat around the kidneys. If there is time, and the animal is not in a starving condition, the Grizzly Bear always runs at the sight of man ; but should the hunter come too suddenly on him, the fierce beast always commences the engagement.—And the first shot of the hunter is a matter of much importance, as, if unsuccessful, his next move must be to look for a sapling to climb for safety. It is rare to find a man who would willingly come into immediate contact with one of these powerful and vindictive brutes. Some were killed near ' Green Springs,' on the Stanislaus, in the winter of 1849–50, that were nearly eight hundred pounds weight. I saw many cubs at San Francisco, Sacramento city, and Stockton, and even those not larger than an ordinary sized dog, showed evidence of their future fierceness, as it required great patience to render them gentle enough to be handled with impunity as pets. In camping at night, my friend ROBERT LAYTON, and I too, often thought what sort of defence we could make should an old fellow come smelling round our solitary tent for supper ; but as ' Old Riley,' our pack-mule, was always tied near, we used to quiet ourselves with the idea that while Riley was snorting and kicking, we might place a couple of well aimed balls from our old friend Miss Betsey (as the boys had christened my large gun), so that our revolvers, COLT's dragoon pistols, would give us the victory ; but really a startling effect would be produced by the snout of a Grizzly Bear being thrust into your tent, and your awaking at the noise of the sniff he might take to induce his appetite.

" I was anxious to purchase a few of the beautiful skins of this species, but those who had killed ' an old Grizzly,' said they would take his skin *home.* It makes a first rate bed under the thin and worn blanket of the digger.

" The different colours of the pelage of this animal, but for the uniformity of its extraordinary claws, would puzzle any one not acquainted with its form, for it varies from jet black in the young of the first and second winter to the hoary gray of age, or of summer."

In TOWNSEND'S "Narrative of a Journey across the Rocky Mountains to the Columbia River, &c." (Philadelphia, 1839), we find two adventures with the Grizzly Bear. The first is as follows : The party were on Black Foot river, a small stagnant stream which runs in a northwesterly direction down a valley covered with quagmires through which they had great difficulty in making their way. "As we approached our encampment, near a small grove of willows on the margin of the river, a tremendous Grizzly Bear rushed out upon us. Our horses ran wildly in every direction, snorting with terror, and became nearly unmanageable. Several balls were instantly fired into him, but they only seemed to increase his fury. After spending a moment in rending each wound (their invariable practice), he selected the person who happened to be nearest, and darted after him, but before he proceeded far he was sure to be stopped again by a ball from another quarter. In this way he was driven about amongst us for perhaps fifteen minutes, at times so near some of the horses that he received several severe kicks from them. One of the pack-horses was fastened upon by the brute, and in the terrified animal's efforts to escape the dreaded gripe, the pack and saddle were broken to pieces and disengaged. One of our mules also lent him a kick in the head, while pursuing it up an adjacent hill, which sent him rolling to the bottom. Here he was finally brought to a stand. The poor animal was so completely surrounded by enemies that he became bewildered. He raised himself upon his hind feet, standing almost erect, his mouth partly open, and from his protruding tongue the blood fell fast in drops. While in this position he received about six more balls, each of which made him reel. At last, as in complete desperation, he dashed into the water, and swam several yards with astonishing strength and agility, the guns cracking at him constantly. But he was not to proceed far. Just then, RICHARDSON, who had been absent, rode up, and fixing his deadly aim upon him, fired a ball into the back of his head, which killed him instantly. The strength of four men was required to drag the ferocious brute from the water, and upon examining his body he was found completely riddled ; there did not appear to be four inches of his shaggy person, from the hips upward, that had not received a ball. There must have been at least thirty shots made at him, and probably few missed him, yet such was his tenacity of life that I have no doubt he would have succeeded in crossing the river, but for the last shot in the brain. He would probably weigh, at the least, six hundred pounds, and was about the height of an ordinary steer. The spread of the foot, laterally, was ten inches, and the claws measured seven inches in length. This animal was remarkably lean ; when in good condition he would doubtless much exceed in weight the estimate I have given."

At p. 68, TOWNSEND says : "In the afternoon one of our men had a somewhat perilous adventure with a Grizzly Bear. He saw the animal crouching his huge frame in some willows which skirted the river, and approaching him on horseback to within twenty yards, fired upon him. The Bear was only slightly wounded by the shot, and with a fierce growl of angry malignity, rushed from his cover, and gave chase. The horse happened to be a slow one, and for the distance of half a mile the race was hard contested, the Bear frequently approaching so near the terrified animal as to snap at his heels, whilst the equally terrified rider, who had lost his hat at the start, used whip and spur with the most frantic diligence, frequently looking behind, from an influence which he could not resist, at his rugged and determined foe, and shrieking in an agony of fear, 'shoot him! shoot him!' The man, who was one of the greenhorns, happened to be about a mile behind the main body, either from the indolence of his horse or his own carelessness; but as he approached the party in his desperate flight, and his lugubrious cries reached the ears of the men in front, about a dozen of them rode to his assistance, and soon succeeded in diverting the attention of his pertinacious foe. After he had received the contents of all the guns, he fell, and was soon despatched. The man rode in among his fellows, pale and haggard from overwrought feelings, and was probably effectually cured of a propensity for meddling with Grizzly Bears."

GEOGRAPHICAL DISTRIBUTION.

The Grizzly Bear has been found as far north as about latitude 61°. It is an inhabitant of the western and northwestern portions of North America, is most frequently met with in hilly and woody districts, and (east of the Rocky Mountains) along the edges of the Upper Missouri and Upper Mississippi rivers, and their tributaries. On the west coast it is found rather numerously in California, generally keeping among the oaks and pines, on the acorns and cones of which it feeds with avidity.

The Grizzly Bear does not appear to have been seen in eastern Texas or the southern parts of New Mexico, and as far as we have heard has not been discovered in Lower California.

GENERAL REMARKS.

To LEWIS and CLARK we are indebted for the first authentic account of the difference between this species and the Black Bear of America, although the Grizzly Bear was mentioned a long time previously by LA HONTAN and others.

DE WITT CLINTON, in a discourse before the New York Literary and Philosophical Society, was the next naturalist who clearly showed that this animal was specifically distinct from either the Polar or the common Bear.

LEWIS and CLARK's name, Grizzly, translated into *Ferox*, has been generally adopted by naturalists to designate this species, and we have admitted it in our nomenclature of this work. We believe that the name proposed for it by ORD (*Ursus horribilis*), and which SAY adopted, must, if we adhere to the rules by which naturalists should be guided in such matters, ultimately take the precedence.

The difference between the Grizzly Bear and the Black may be easily detected. The soles of the feet of the former are longer, and the heel broader; the claws are very long, whilst in the Black Bear they are quite short. The tail of the Grizzly Bear is shorter than that of the Black, and its body is larger, less clumsy and unwieldy, and its head flatter than the head of the latter.

The Grizzly Bear makes enormous long tracks, and differs widely from the Black Bear in its habits, being very ferocious, and fearlessly attacking man.

We think the average size and weight of this animal are much under-rated. We have no hesitation in stating that the largest specimens would weigh considerably over one thousand pounds. We have seen a skin of the common Black Bear, shot in the State of New York, the original owner of which was said to have weighed twelve hundred and odd pounds when killed!

Plate CXXXII

Drawn from Nature by J.W. Audubon

On stone by W.E. Hitchcock

Lith Printed & Col.d by J.T. Bowen, Philad.a

Hare Indian Dog

CANIS FAMILIARIS.—LINN. (VAR. LAGOPUS.)

HARE-INDIAN DOG.

PLATE CXXXII.—MALE.

C. Magnitudine inter lupum et vulpem fulvum intermedius, auriculis erectis, cauda comosa, colore cinereo, albo nigroque notato.

CHARACTERS.

Intermediate in size between the wolf and red fox ; ears, erect ; tail, bushy ; colour, gray, varied with white and dark markings.

DESCRIPTION.

The Hare-Indian Dog resembles the wolf rather more than the fox. Its head is small, muzzle slender, ears erect, eyes somewhat oblique, legs slender, feet broad and hairy, and its tail bushy and generally curled over its hip. The body is covered with long hair, particularly about the shoulders. At the roots of the hair, both on the body and tail, there is a thick wool. On the posterior parts of the cheeks the hair is long and directed backwards, giving the animal the appearance of having a ruff around the neck.

COLOUR.

Face, muzzle, belly, and legs, cream white ; a white central line passes over the crown of the head to the occiput ; the anterior surface of the ear is white, the posterior yellowish-gray or fawn colour ; tip of nose, eye-lashes, roof of mouth, and part of the gums, black ; there is a dark patch over the eye, and large patches of dark blackish-gray or lead colour, on the body mixed with fawn colour and white, not definite in form, but running into each other. The tail is white beneath, and is tipped with white.

DIMENSIONS.

	Feet.	Inches.
Length of head and body, about - - - -	3	
Height at shoulder, about - - - - -	1	2
Length of tail, - - - - - - -	1	3

HABITS.

This animal is more domestic than many of the wolf-like Dogs of the plains, and seems to have been entirely subjugated by the Indians north of the great lakes, who use it in hunting, but not as a beast for burthen or draught.

Sir John Richardson says (F. B. A., p. 79): "The Hare-Indian Dog is very playful, has an affectionate disposition, and is soon gained by kindness. It is not, however, very docile, and dislikes confinement of every kind. It is very fond of being caressed, rubs its back against the hand like a cat, and soon makes an acquaintance with a stranger. Like a wild animal it is very mindful of an injury, nor does it, like a spaniel, crouch under the lash; but if it is conscious of having deserved punishment, it will hover round the tent of its master the whole day, without coming within his reach, even if he calls it. Its howl, when hurt or afraid, is that of the wolf; but when it sees any unusual object it makes a singular attempt at barking, commencing by a kind of growl, which is not, however, unpleasant, and ending in a prolonged howl. Its voice is very much like that of the prairie wolf.

"The larger Dogs which we had for draught at Fort Franklin, and which were of the mongrel breed in common use at the fur posts, used to pursue the Hare-Indian Dogs for the purpose of devouring them; but the latter far outstripped them in speed, and easily made their escape. A young puppy, which I purchased from the Hare Indians, became greatly attached to me, and when about seven months old ran on the snow by the side of my sledge for nine hundred miles, without suffering from fatigue. During this march it frequently of its own accord carried a small twig or one of my mittens for a mile or two; but although very gentle in its manners it showed little aptitude in learning any of the arts which the Newfoundland Dogs so speedily acquire, of fetching and carrying when ordered. This Dog was killed and eaten by an Indian, on the Saskatchewan, who pretended that he mistook it for a fox."

The most extraordinary circumstance in this relation is the great endurance of the puppy, which certainly deserves special notice. Even the oldest and strongest Dogs are generally incapable of so long a journey as nine hundred miles (with probably but little food), without suffering from fatigue.

GEOGRAPHICAL DISTRIBUTION.

It is stated by Sir John Richardson that this species exists only among the different tribes of Indians that frequent the borders of Great Bear lake and the Mackenzie river.

GENERAL REMARKS.

From the size of this animal it might be supposed by those who are desirous of tracing all the Dogs to some neighbouring wolf, hyena, jackal, or fox, that it had its origin either from the prairie wolf or the red fox, or a mixture of both.

The fact, however, that these wolves and foxes never associate with each other in the same vicinity, and never have produced an intermediate variety, or, that we are aware of, have ever produced a hybrid in their wild state, and more especially the fact that the prairie wolf, as stated by RICHARDSON, does not exist within hundreds of miles of the region where this Dog is bred, must lead us to look to some other source for its origin.

Its habits, the manner in which it carries its tail, its colour, and its bark, all differ widely from those of the prairie wolf.

We have never had an opportunity of seeing this animal and examining it, except in the stuffed specimen from which our drawing was made ; we are therefore indebted to Sir JOHN RICHARDSON for all the information we possess in regard to its habits, and have in this article given the results of his investigations mostly in his own language.

LEPUS TEXIANUS.—Aud. and Bach.

Texan Hare—Vulgo Jackass Rabbit.

PLATE CXXXIII.—Male.

L. Magnitudine, L. Californicum excedens, auriculis maximis, capite tertia parte longioribus, linea fusca supra in collo, striâ nigrâ a natibus usque ad caudæ apicem productâ, corpore supra luteo nigroque vario, subter, collo rufo gula atque ventre albis.

CHARACTERS.

Larger than the Californian Hare ; ears, very large—more than one third longer than the head ; a dark brown stripe on the top of the neck, and a black stripe from the rump, extending to the root of the tail and along its upper surface to the tip. Upper surface of body, mottled deep buff and black, throat and belly white, under side of neck dull rufous.

DESCRIPTION.

Crown of the head, depressed or flattened, forming an obtuse angle with the forehead and nose ; ears, of immense size, being larger than in any other species of Hare known to us. Body, full, and rather stout ; forelegs, of moderate length and size ; thighs, stout and large ; tarsus, of moderate length ; nails, strong, deeply channelled beneath.

COLOUR.

Hairs on the upper surface of body, white from the roots for two thirds of their length, then brown, then dull buff, and tipped very narrowly with black. On the belly, throat, and insides of legs, the hairs are white from the roots to the tips.

One of our specimens has a black patch on the inner surface of the ear near its base ; another has a brown patch in that place ; anterior margin of the ears, buff ; posterior portion of the ear for an inch and a half from the tip, whitish ; a narrow line of dark brown runs from between the ears for an inch along the back of the neck ; the anterior outer half of the ear, and the posterior inner half of the ear, are clothed with a mixture of parti-

Plate LXXXIII.

Drawn from Nature by J. W. Audubon

On stone by W. E. Hitchcock

Lith Printed & Col'd by J. T. Bowen, Philad.

Texian Hare.

coloured gray and yellowish hairs ; the posterior outside half of the ear is white, with the exception of the extreme point, which in one of our specimens has a slight margin of brown at the tip of the ear, while another specimen is more deeply tinged with brown for three fourths of its length.

Around the eye there is a light yellowish-gray ring ; under surface of neck, rufous, faintly spotted or marked with brown ; tail, black above, the same colour continuing on the rump and dorsal line in a stripe for about four inches from the root of the tail ; eyes, orange hazel ; nails, brown. The line of white on the belly and flanks is irregular in shape where it joins the dark colours of the upper surface, and in this respect differs from *Lepus callotis*, in which species the white extends higher up the sides and is continued in a tolerably straight line nearly to the tail.

Whiskers, white, a few of them black at the roots.

DIMENSIONS.

	Feet.	Inches.	Lines.
From point of nose to root of tail, -	1	9	
" " to ear,		4	1
Ear, externally, -		6	5
Width of ear, -		3	
Length of tarsus, -		5	
" tail (including fur),		4	2
" longest whisker, -		3	6

HABITS.

This Hare received from the Texans, and from our troops in the Mexican war, the name of Jackass rabbit, in common with *Lepus callotis*, the Black-tailed Hare described in our second volume, p. 95. It is the largest of three nearly allied species of Hare which inhabit respectively New Mexico, Texas, Mexico, and California, viz. the present species, the Black-tailed, and the Californian Hare. It is quite as swift of foot as either of the others, and its habits resemble those of the Black-tailed Hare in almost every particular. The young have generally a white spot on the middle of the top of the head, and are remarkable for the rigidity of the fringe of hairs which margins the ears. The feet of this species do not exhibit the red and dense fur which prevails on the feet of the Black-tailed Hare (and from which it has sometimes been called the Red-footed Hare).

The Mexicans are very fond of the flesh of this animal, and as it is widely distributed, a great many are shot and snared by them. It is very good eating, and formed an important item in the provisions of JOHN W. AUDU-

BON'S party whilst passing through Mexico, they at times killing so many that the men became tired of them.

Fabulous stories similar to those related of many other animals of which little was formerly known, have been told us of this Hare, which has been described as enormously large, and was many years ago mentioned to us as equal in size to a fox. Of course we were somewhat disappointed when we procured specimens, although it is a fine large species.

Among other old stories about the animals of Texas and Mexico, we have a rather curious one in CLAVIGHERO's notes or attempted elucidation of HERNANDEZ, which we give as translated by Capt. J. P. MCCOWN from the Spanish. The *Ocotochtli*, according to Dr. HERNANDEZ, is a species of wild-cat. He says that " when it has killed any game it climbs a tree and utters a howl of invitation to other animals that come and eat and die, as the flesh was poisoned by its bite, when he descends and makes his meal from the store that his trick has put at his disposal."

GEOGRAPHICAL DISTRIBUTION.

This Hare appears to inhabit the southern parts of New Mexico, the western parts of Texas, and the elevated lands westward of the *tierras calientes* (low lands of the coast) of Mexico, and is found within a few miles of San Petruchio, forty miles from the coast: so J. W. AUDUBON was informed by some Rangers who accompanied a party sent from San Antonio in 1845, who having the use of " Col. HARNEY'S" greyhounds, had many a chase, but never caught one! How near it approaches the sea coast we could not learn. It was not observed west of Ures in Sonora by J. W. AUDUBON, and seems to be replaced by the Californian Hare on the Pacific coast.

Its southern limit is unknown to us, but it probably extends some distance beyond the city of Mexico.

GENERAL REMARKS.

Since publishing our article on *Lepus Townsendii* we have received some accounts of the habits of a Hare which we presume may prove to be that animal; they are singular, and may interest our readers. Captain THOMAS G. RHETT, of the United States army, who was stationed at Fort Laramie for more than two years, observed the Hares of that neighbourhood to make burrows in the ground like rabbits. They ran into these holes when alarmed, and when chased by his greyhounds generally escaped by diving into them. The captain frequently saw them sitting at the mouths of their holes like prairie dogs, and shot them. Several that he thus killed had only their heads exposed outside of their burrow.

These holes or burrows are dug in a slanting direction, and not straight up and down like the badger holes. The females bring forth their young in them, and their habits must assimilate to those of the European rabbit. The captain states that they turn white in winter, but as he made no notes and brought no specimens, we cannot with certainty decide that they were the animal we named *L. Townsendii*. Should they prove to be the same, however, the name will have to be changed to *L. campestris*, a Hare of the plains which we had previously described, but subsequently thought was not that species, as it became white in winter, which we were told *L. Townsendii* did not. See our first volume, p. 30.

ARCTOMYS FLAVIVENTER.—Bach.

YELLOW-BELLIED MARMOT.

PLATE CXXXIV.—Male.

A. Supra flavido-albo nigroque griseus, capitis vertice nigro, subtus saturate flavus, nasi extremitate labiis, mentoque albis, pedibus fuscescente flavis, cauda subnigra.

CHARACTERS.

Upper parts, grizzled yellowish-white and black ; crown of the head, chiefly black ; under parts, deep yellow ; point of nose, lips, and chin, white ; feet, brownish-yellow ; tail, blackish-brown.

SYNONYMES.

Arctomys Flaviventer. Bachman, Proc. Acad. Nat. Sci. Phila., October 5, 1841.
" " Catal. Zool. Soc. 1839, Specimen No. 459, Bachman's MSS

DESCRIPTION.

In form this animal resembles the figures and descriptions of what was formerly considered the Canada Marmot (*Arctomys empetra*), which has since been ascertained to be the young of the Maryland Marmot (*A. monax*).

Head, rather small ; ears, small and narrow ; nails, short ; tail, rounded, and rather long ; the whole animal is thickly clothed with fur, somewhat softer than that of the Maryland Marmot.

The upper incisors have several indistinct longitudinal grooves.

COLOUR.

Fur on the back, grayish-black at base ; on each hair a considerable space is occupied by dirty yellowish-white, which is gradually shaded towards the tips through brown into black, but the tips are yellowish-white.

Hairs on the under surface, grayish-black at base ; hairs of the feet, chiefly black at base ; cheeks, grizzled with white and dark brown, the latter colour prevailing ; a rusty brown patch on the throat borders the

Plate CXXXIV.

Drawn from Nature by J.W.Audubon.

On stone by W.E.Hitchcock.

Lith Printed & Col. by J.T. Bowen, Philad.ª

Yellow bellied Marmot

white hairs on the chin; whiskers, mostly black; palms, entirely naked through their whole extent. There is an indistinct yellow elongated spot behind the nose, and also one behind or above the eye.

DIMENSIONS.

	Inches.	Lines
From point of nose to root of tail, - - -	16	
Tail, to end of fur, - - - - - -	6	10
Heel, to point of nail, - - - - -	2	6½
Height of ear posteriorly, - - - - -		6½
From point of nose to ear, - - - -	3	

HABITS.

The specimen from which our description of this Marmot was drawn up, was found by us among the skins sent to England by DRUMMOND and DOUGLAS, procured by those gentlemen in our northwestern territories, and placed in the museum of the Zoological Society of London. Since we described it, the skin has been stuffed and set up.

Not a line was written in regard to its habits or the place where it was killed; its form and claws, however, indicate that like the other species of Marmot found in America, it is a burrowing animal, and feeds on seeds, roots, and grasses. We may also presume it has four or five young at a birth.

GEOGRAPHICAL DISTRIBUTION.

As just stated, the exact locality in which this animal was captured has not been given, but judging from the route travelled over by DOUGLAS, we presume it was obtained in the mountainous districts that extend north and south between Western Texas and California, where it probably exists, but if seen has been supposed by the hunters and miners to be the common Marmot or woodchuck of the Atlantic States (*A. monax*).

GENERAL REMARKS.

This species differs from the young of *Arctomys monax*, by some naturalists named *A. empetra*, as we ascertained by comparing it with several specimens of that so-called species, in the museum of the Zoological Society, its feet being yellow instead of black, as in those specimens, and the belly yellow, not deep rusty red. Besides, the hairs on the back are yellowish-white and black, in place of rusty brown, black, and white.

The head is narrower, the toes smaller, and the claws only half as long, as in the above specimens. The ears are also considerably smaller, narrower and more ovate than the ears of *A. monax*, which are round.

Plate CXXXV.

N°.27

Drawn from Nature by J.W.Audubon

On stone by W.E.Hitchcock.

Lith. Printed & Col.d by J.T.Bowen, Philad.a

Richardson's Meadow Mouse

ARVICOLA RICHARDSONII.—Aᴜᴅ. and Bᴀᴄʜ

PLATE CXXXV. Fɪɢ. 1.

A. fuscus nigro tinctus, subtus cinereus, cærulescente-canus, auriculis mediocribus vellere fere conditis, cauda capite paullulum longiore.

CHARACTERS.

Dull brown mixed with black, under parts bluish-gray ; ears, of moderate size, nearly hidden by the fur ; tail, a little longer than the head.

SYNONYME.

Aʀᴠɪᴄᴏʟᴀ Rɪᴘᴀʀɪᴜs? Ord. Bᴀɴᴋ Mᴇᴀᴅᴏᴡ-Moᴜsᴇ. Richardson, F. B. A., p. 120.

DESCRIPTION.

Head, rather large : incisors, large, much exposed, and projecting beyond the nose—upper, flattened anteriorly, marked with scarcely perceptible perpendicular grooves, and with a somewhat irregular and rather oblique cutting edge—lower, twice as long as the upper, and narrower, slightly curved, and rounded anteriorly ; nose, thick and obtuse : whiskers, few and rather short ; eyes, rather small ; ears, ovate, rounded at the tip, not easily distinguishable until the surrounding fur is blown or moved aside.

Body, more slender behind than at the shoulders, the hind-legs not being so far apart as the fore-legs ; tail, rather short, tapering, and thinly covered with short hairs ; fore-legs, short ; feet, rather small, with four slender, well separated toes, and the rudiment of a thumb, which is armed with a minute nail ; claws, small, compressed, and pointed ; the third toe nearly equals the middle one, which is the longest.

The hair of the toes projects over the claws but does not conceal them ; the toes of the hind-feet are longer than those of the fore-feet, and their claws are somewhat longer ; the inner one is the shortest, the second longer than the third, and the third longer than the fourth ; the first and fifth are considerably shorter than the others, and are placed farther back.

The fur on the back is about eight lines long, but not so soft and fine as in some other animals of the genus ; it is nearly as long on the crown and cheeks, but is shorter and thinner on the chest and belly.

COLOUR.

Incisors, yellow ; claws, white ; whiskers, black ; the whole dorsal aspect, including the shoulders and outsides of the thighs, is dull or dusky brown, proceeding from an intimate mixture of yellowish-brown and black, which colours are confined to the tips of the hairs and are so mingled as to produce a nearly uniform shade of colour without lustre.

From the roots to near the tips, the fur has a uniform shining blackish-gray colour ; on the ventral aspect (lower parts) it is bluish-gray ; the margin of the upper lip, the chin, and the feet, are dull white ; tail, dark brown above, lighter beneath, the two colours meeting by an even line.

DIMENSIONS.

Length of head and body, - - - - - -	7 inches.
" tail, - - - - - - - -	2 "

HABITS.

DRUMMOND, who procured this Meadow-Mouse, states that its habits are analogous to those of the common water-rat of Europe (*Arvicola amphibius*), with which it may be easily confounded, although the shortness of its tail may serve as a mark by which to distinguish it.

It frequents moist meadows amongst the Rocky Mountains, and swims and dives well, taking to the water at once when pursued. All Meadow-Mice indeed are capital swimmers. We some time since amused ourselves watching one that had fallen into a circular cistern partly built up with stone and partly excavated out of the solid rock by blasting, and which was plastered with cement on the inside to make it water-tight. This cistern had about four feet of water in it. On one side there was a projecting rounded knob of stone some five or six inches long and about two wide, which slanted out of the water so that the upper edge of it was dry. Upon this little resting-place there was a large *Arvicola Pennsylvanica* (Wilson's Meadow-Mouse) seated very quietly, having probably tumbled in the preceding night. When we approached the edge and looked down into the clear element we at first did not observe the Rat, but as soon as we espied him he saw us, immediately dived, and swam around underneath the surface quite rapidly ; he soon arose, however, and regained his position on the ledge, and we determined to save him from what had been his impending fate—drowning or starving, or both. We procured a plank, and gently lowering one end of it towards the ledge, thought he would take advantage of the inclined plane thus afforded him, to come out : but in our awkwardness we suffered the plank to slip, and at the plash in the water the little fellow dived and swam around several times before he again returned to his resting place, where we now had the end of the board fixed, so that he could get upon it. As soon as he was on it, we began to raise the plank, but when we had him about three feet above the surface he dashed off into the water, making as pretty a dive as need be. He always looked quite dry, and not a hair of his coat was soiled or turned during these frequent immersions, and it was quite interesting to see the inquisitive looks he cast towards us, turning his head and appearing to have strong doubts whether we meant to help, or to make an end of him. We put down the plank again, and after two attempts, in both of which his timidity induced him to jump off it when he was nearly at the edge of the cistern, he at last reached the top, and in a moment disappeared amid the weeds and grasses around.

GEOGRAPHICAL DISTRIBUTION.

The only information we possess of the habitat of this animal is from DRUMMOND, who states that he captured it near the foot of the Rocky Mountains.

GENERAL REMARKS.

This species possesses longer and stronger incisors than any other American Rat of this genus ; its mouth presenting in fact a miniature resemblance to that of the musk-rat.

Although the *Arvicola xanthognatha* is a larger animal than the present, yet its incisors are not more than half as long as in this species.

We have named this Arvicola in honour of Sir JOHN RICHARDSON, who in describing it (Fauna Boreali Americana, p. 120), applied to it, with a doubt, the name of *Arvicola riparius*, ORD, from which it differs so much as to render a comparison here unnecessary.

ARVICOLA DRUMMONDII.—Aud. and Bach

DRUMMOND'S MEADOW-MOUSE.

PLATE CXXXV. Figure with Short Tail.—Summer pelage.

A. Corpore supra fusco, infra fusco-cinereo, ad latera rufo tíncto, robustiore et paulo majore quam in A. Pennsylvanicâ ; auriculis vellere fere occultis ; cauda brevi, capitis dimidium subequante.

CHARACTERS.

Body, above, dark brown ; beneath, dull brownish-gray tinged with red. Stouter and rather larger than Wilson's Meadow-Mouse (A. Pennsylvanica) ; *ears, scarcely visible beyond the fur ; tail, short, about half the length of the head.*

SYNONYMES.

ARVICOLA NOVEBORACENSIS—SHARP-NOSED MEADOW-MOUSE. Rich., F. B. A., p. 126.

DESCRIPTION.

Body, thick ; head, of moderate size, tapering from the ears to the nose ; nose, slender and more acute than in many other *Arvicolæ*, projecting a little beyond the incisors, which are rather large.

Ears, rounded, scarcely visible beyond the fur ; tail, covered with short hairs, scarcely concealing the scales, converging to a point at the tip ; legs, very short ; feet, rather small ; claws, weak and compressed ; a very minute nail occupies the place of the thumb ; the fur is a little coarser than that of *A. Pennsylvanica.*

The whiskers, which are not numerous, reach the cheeks.

COLOUR.

Hair on the back, and upper part of the head, grayish-black from the roots to near the tips, which are reddish-brown terminated with black ; the resulting colour is an intimate mixture of brown and black, appearing in some lights dark reddish-brown, in others yellowish-brown mixed with blackish ; around the eyes, yellowish-red ; there is a lightish space behind the ears and along the sides ; under surface, yellowish-gray, mingling on the sides with the colour of the back ; upper surface of the tail, dark brown ; under side, grayish-white ; feet, dark gray, tinged with rufous.

DIMENSIONS.

		Inches.	Lines.
Length of head and body, - - - - -		4	3
" head, - - - - - - -		1	4
" tail, - - - - - - -		1	

HABITS.

The specimen from which our drawing was made is one of those obtained by Mr. DRUMMOND, and was deposited by that gentleman in the museum of the Zoological Society at London, as well as many others to which we have already referred in our work. It was examined and described by Sir JOHN RICHARDSON, who mistook the animal for a supposed species found in the state of New York, and loosely described by RAFFINESQUE under the name of *Lemmus noveboracensis*, and which we refer to *A. Pennsylvanica*, with which we have compared the description.

DRUMMOND in regard to the habits of the present animal merely states that they are similar to those of *Arvicola xanthognatha*.

GEOGRAPHICAL DISTRIBUTION.

Valleys of the Rocky Mountains.

GENERAL REMARKS.

As above mentioned, Sir JOHN RICHARDSON described this animal, quoting from DESMAREST (Mamm., p. 286), RAFFINESQUE'S description of the so-called *Lemmus noveboracensis*, which appears to apply to one of the varieties of WILSON's Meadow-Mouse (*Arvicola Pennsylvanica*), of which we possess specimens.

From an examination of many species, we have arrived at the conclusion that no *Arvicolæ* found on the Rocky Mountains are identical with any in the Atlantic States, and on a comparison of RICHARDSON'S species with those referred to by RAFFINESQUE, we determined without much hesitation that the present is a new species under an old name, and we have consequently attached to it the name of its discoverer—DRUMMOND.

By some oversight this species was not named on our plate as distinct from *A. Richardsonii*, but is easily distinguished by its short tail—the two being on the same engraving.

CERVUS VIRGINIANUS.—Pennant.

Common American Deer.

PLATE CXXXVI.—Male and Female.

(Fawn.) PLATE LXXXI.—Winter Pelage.

In our article on the Virginian Deer (vol. ii. p. 220), we gave descriptions of the characters and habits of this species; we now present figures of the adult male and female.

We have not much information to add to that already given: it may be of interest, however, to notice the annual changes which take place in the growth of the horns, from adolescence to maturity, and the decline which is the result of age.

At Hyde Park, on the estate of J. R. Stuyvesant, Esq., Dutchess county, New York, seven or eight Deer were kept for many years, and several raised annually. We had the opportunity at the hospitable mansion of Mr. Stuyvesant, of examining a series of horns, all taken from the same buck as they were annually shed, from the first spikes to the antlers that crowned his head when killed; and we now give a short memorandum showing the progress of their growth from year to year. In 1842, when this buck was one year old, his horns (spikes) had each one rudimentary prong—one about five eighths of an inch long, the other scarcely visible; in 1843 they had two prongs four to six inches long; in 1844, three prongs, and brow antlers, longest prong eight inches; in 1845, a little larger in diameter, brow antlers longer and curved; 1846, rather less throughout in size; 1847, the two last prongs quite shortened. These last were somewhat broken by an accident, but evidently show that the animal had lost some degree of vigour. Age when killed, six years.

It should be observed that this animal was restricted to a park and was partially domesticated, being occasionally fed a little in the winter season; and being thus deprived of the wider range of the forest, the horns may not have exhibited all the peculiarities of the wild unrestrained buck.

We think however that the above will give a tolerably correct idea of the operations of nature in the annual production and conformation of the horns. They become longer and more branched for several years, until the animal has arrived at maturity, when either from age or disease they begin to decline.

In connection with this subject it may not.be uninteresting to notice the

Plate CXXXVI

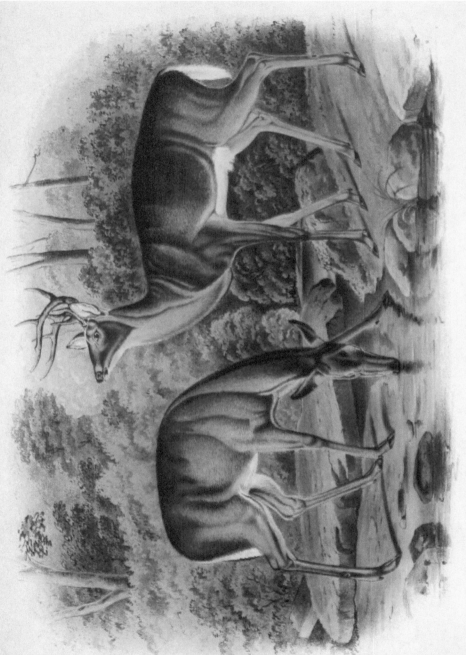

Drawn from Nature by J. W. Audubon

On Stone by W™ E. Hitchcock

Common or Virginian Deer

Lith Printed & Col. by J. T. Bowen Phil.

effect of castration on the horns of the buck. When this operation has been performed during the season when the horns are fully grown, it is said they are not dropped, but continue on the head for many years ; when the operation has been performed after they are dropped, there is no subsequent growth of horns, and the head appears ever afterwards like that of a doe.

We had an opportunity at the Blue Sulphur Springs in Virginia, of examining two tame bucks which had been castrated during the time that their horns were in velvet. Their horns continued to grow for several years ; the antlers were of enormous length, and very irregularly branched, but the velvet was still retained on them ; they presented a soft spongy appearance, and from slight scratches or injuries were continually bleeding ; the neck had ceased to swell periodically as in the perfect bucks, they had become very large, seemed to be quite fat, and when first seen at a distance we supposed them to be elks.

GENUS ENHYDRA.—Fleming.

DENTAL FORMULA.

Incisive $\frac{6}{4}$; *Canine* $\frac{4-4}{5-5}$; *Molar* $\frac{2-2}{3-3} = 38$.

Head, small and globular; ears, short and conical, placed far back in the head.

Body, very long, covered with a dense glossy fur; tail, less than one fourth the length of the body, rather stout, depressed, covered with strong hairs on the sides.

Hind-feet, webbed.

LICHTENSTEIN says this genus has hind-feet like those of the common seals, ears resembling those of the seals of the genus *Otaria*, and a tail similar to that of the common Otter.

He places the Sea Otter (correctly, as we think) between the Otter and the seals that possess ears (*Otaria*).

Mammæ, two—ventral.

There is only one species in the genus.

Habit, living principally at sea and in bays and estuaries.

The generic name is derived from ενυδρος, *enudros*, aquatic; Gr. εν, *en*, in, and ὑδωρ, *hudōr*, water.

ENHYDRA MARINA.—Erxleben.

SEA OTTER.

PLATE CXXXVII.—Male.

E. perelongata, cauda depressa, corporis partem quartam æquante, pedibus posticis curtis, istis Phocarum similibus, colore castaneo vel nigro, vellere mollissimo; Lutrâ Canadensis duplo major.

CHARACTERS.

Body, very much elongated; tail, depressed, and one fourth the length of the body; hind-feet, short, and resembling those of the seal; colour, chesnut brown or black; twice the size of the common Otter; fur, exceedingly fine.

Plate CXXXVII

Drawn from Nature by J. W. Audubon.

Lith. Printed & Col.ᵈ by J. T. Bowen, Philad.ᵃ

Sea Otter

SYNONYMES.

MUSTELA LUTRIS. Linn.
SEA BEAVER. Krascheninikoff, Hist. Kamsk. (Grieve's Trans.), p. 131.   Ann. 1764.
MUSTELA LUTRIS. Schreber, Saügethiere, p. 465, fig. t. 128.
LUTRA MARINA. Erxleben, Syst.   Ann. 1777.
   "   "   Steller, Nov. Com. Petrop., vol. ii. p. 267, t. 16.
SEA OTTER. Cook's Third Voyage, vol. ii. p. 295.   Ann. 1784.
   "   "   Pennant's Arctic Zoology, vol. i. p. 88.   Ann. 1784.
LUTRA STELLERI. Lesson, Manual, pp. 156, 423.
SEA OTTER. Meares, Voyage, pp. 241, 260.   Ann. 1790.
   "   "   Menzies, Philos. Trans., p. 385.   Ann. 1796.
ENHYDRA MARINA. Fleming, Phil. Zool., vol. ii. p. 187.   Ann. 1822.
ENYDRIS STELLERI. Fischer, Synopsis, p. 228.
LUTRA MARINA. Harlan, Fauna, p. 72.
THE SEA OTTER. Godman's Nat. Hist., vol. i. p. 228.
ENYDRIS MARINA. Licht., Darstellung neuer oder wenig bekannter Saügethiere.
         Berlin, 1827–1834.   Tafel xlix.
LUTRA (ENHYDRA) MARINA. Rich., Fauna Boreali Americana, p. 59.

DESCRIPTION.

Head, small in proportion to the size of the body ; ears, short, conical, and covered with hair ; eyes, rather large ; lips, thick ; mouth, wide, and furnished with strong and rather large teeth ; fore-feet, webbed nearly to the nails, and much like those of the common Otter, five claws on each. Hind-legs and thighs, short, and better adapted for swimming than in other mammalia except the seals ; hind-feet, flat and webbed, the toes being connected by a strong granulated membrane, with a skin skirting the outward toe ; all the webs of the feet are thickly clothed with glossy hairs about a line in length.

One of the specimens referred to by Mr. MENZIES (the account of which is published in the Philosophical Transactions) measured eight inches across the hind-foot ; the tongue was four inches long and rounded at the end, with a slight fissure, giving the tip a bifid appearance.

The tail is short, broad, depressed, and pointed at the end ; the hair both on the body and tail is of two kinds—the longer hairs are silky, glossy, and not very numerous, the fur or shorter hair exceedingly soft and fine.

COLOUR.

The cheeks generally present a cast of grayish or silvery colour, which extends along the sides and under the throat ; there is a lightish circle around the eye ; top of the head, dark brown ; the remainder of the body (above and beneath) is deep glossy brownish-black.

There is a considerable variety of shades in different specimens, some being much lighter than others. The longer hairs intermixed with the fur are in the best skins black and shining. In some individuals the fur about the ears, nose, and eyes is either brown or light coloured ; the young are sometimes very light in colour, with white about the nose, eyes, and forehead.

The fur of the young is not equal in fineness to that of the adult.

DIMENSIONS.

Adult.

	Feet.	Inches.
Length from point of nose to root of tail, - -	4	2
" of tail, - - - - - - -	1	

Young, about two years old.

	Feet.	Inches.
Length from end of nose to root of tail, - -	3	
" of tail, - - - - - -		7½
Width of head between the ears, - - - -		4
Height of ear, - - - - - - -		¾
From elbow of fore-leg to end of nail, - - -		4½
Length of hind-foot from heel to end of nail, -		6¼
" fore toe, - - - - - -		¼
" inner hind-toe, - - - - -		1
" outer hind-toe, - - - - -		3
Circumference of the head, behind the ears, -		10¼
" of body around the breast, - -	1	5
" " " loins, - -	1	10

HABITS.

Next to the seals the Sea Otter may be ranked as an inhabitant of the great deep : it is at home in the salt waves of the ocean, frequently goes some distance from the " dull tame shore," and is sometimes hunted in sail-boats by the men who live by catching it, even out of sight of land.

But although capable of living almost at sea, this animal chiefly resorts to bays, the neighbourhood of islands near the coast, and tide-water rivers, where it can not only find plenty of food, but shelter or conceal itself as occasion requires.

It is a timid and shy creature, much disconcerted at the approach of danger, and when shot at, if missed, rarely allows the gunner a second chance to kill it.

Hunting the Sea Otter was formerly a favourite pursuit with the few sailors or stray Americans that lived on the shores of the Bay of San Francisco, but the more attractive search for gold drew them off to the mines when SUTTER's mill-race had revealed the glittering riches inter-mixed with its black sands. One of the shallops formerly used for catching the Sea Otter was observed by J. W. AUDUBON at Stockton, and is thus described by him : The boat was about twenty-eight feet long and eight feet broad, clinker built, and sharp at both ends like a whale-boat, which she may in fact have originally been. rigged with two lug sails, and looked like a fast craft. Whilst examining her the captain and owner came up to enquire whether he did not want to send some freight to Hawkins' Bar, but on finding that was not the object of his scrutiny. gave him the follow-ing account of the manner of hunting the Otter.

The boat was manned with four or five hands and a gunner, and sailed about all the bays. and to the islands even thirty or forty miles from the coast, and sometimes north or south three or four hundred miles in quest of these animals. On seeing an Otter the boat was steered quietly for it, sail being taken in to lessen her speed so as to approach gently and without alarming the game. When within short gun-shot, the marksman fires, the men spring to the oars. and the poor Otter is harpooned before it, sinks by the bowsman. Occasionally the animals are sailed up to while they are basking on the banks, and they are sometimes caught in seines. The man who gave this information stated that he had known five Otters to be shot and captured in a day, and he had obtained forty dollars apiece for their skins. At the time J. W. AUDUBON was in California he was asked a hundred dollars for a Sea Otter skin. which high price he attributed to the gold discoveries.

Only one of these Otters was seen by J. W. AUDUBON whilst in Califor-r.ia : it was in the San Joaquin river, where the bulrushes grew thickly on the banks all about. The party were almost startled at the sudden appearance of one, which climbed on to a drift log about a hundred yards above them. Three rifle balls were sent in an instant towards the unsus-pecting creature, one of which striking near it, the alarmed animal slided into the water and sunk without leaving, so far as they could see, a single ripple. It remained below the surface for about a minute, and on coming up raised its head high above the water, and having seen nothing to frighten it, as they judged, began fishing. Its dives were made so gently that it was evidently as much at its ease in the water as a Grebe, and it frequently remained under the surface as long at least as the great northern diver or loon. They watched its movements some time, but could not see that it took a fish, although it dived eight or ten times. On firing another shot, the Otter appeared much frightened (possibly having been touched) and swimming rapidly, without diving, to the opposite shore, disappeared in the rushes, and they did not see it again.

In the accounts of this species given by various authors we find little respecting its habits, and it is much to be regretted that so remarkable an animal should be yet without a full " biography."

Sir JOHN RICHARDSON, who gives an excellent description of its fur from one who was engaged in the trade, says, " It seems to have more the manners of a seal than of the land Otter. It frequents rocks washed by the sea, and brings forth on land, but resides mostly in the water, and is occasionally seen very remote from the shore."

GODMAN states that "its food is various, but principally cuttle-fish, lobsters, and other fish. The Sea Otter, like most other animals which are plentifully supplied with food, is entirely harmless and inoffensive in its manners, and might be charged with stupidity, according to a common mode of judging animals, as it neither offers to defend itself nor to injure those who attack it. But as it runs very swiftly and swims with equal celerity it frequently escapes, and after having gone some distance turns back to look at its pursuers. In doing this it holds a fore-paw over its eyes, much in the manner we see done by persons who in a strong sunshine are desirous to observe a distant object accurately. It has been inferred that the sight of this animal is imperfect ; its sense of smelling, however, is said to be very acute."

The latter part of the above paragraph at least, may be taken as a small specimen of the fabulous tales believed in olden times about animals of which little that was true had been learned.

Dr. GODMAN relates farther that the female Sea Otter brings forth on land after a pregnancy of eight or nine months, and but one at a birth, and states that the extreme tenderness and attachment she displays for her young are much celebrated. According to his account the flesh is eaten by the hunters, but while it is represented by some as being tender, juicy, and flavoured like young lamb, by others it is declared to be hard, insipid, and tough as leather. We advise such of our readers as may wish to decide which of these statements is correct, and who may be so fortunate as to possess the means and leisure, to go to California and taste the animal—provided they can catch or kill one.

We will conclude our very meagre account of the habits of the Sea Otter by quoting the following most sensible remarks from Sir JOHN RICHARDSON, given in a note in the Fauna Boreali Americana, p. 60 : " Not having been on the coasts where the Sea Otter is produced, I can add nothing to its history from my own observation, and I have preferred taking the description of the fur from one who was engaged in the trade, to extracting a scientific account of the animal from systematic works, which are in the hands of every naturalist."

GEOGRAPHICAL DISTRIBUTION.

The Sea Otter inhabits the waters which bound the northern parts of America and Asia, and separate those continents from each other, viz. the North Pacific Ocean and the various seas and bays which exist off either shore from Kamtschatka to the Yellow Sea on the Asiatic side, and from Allaska to California on the American.

GENERAL REMARKS.

Although this animal has been known and hunted for more than a century, and innumerable skins of it have been carried to China (where they formerly brought a very high price), as well as to some parts of Europe, yet no good specimens, and but few perfect skulls of it, exist in any museum or private collection. The difference between the dentition of the young and the adult, being in consequence unknown, has misled many naturalists, and caused difficulties in the formation of the genus.

LINNÆUS, strangely enough, placed it among the martens (*Mustela*); ERXLEBEN, in the genus *Lutra ;* FLEMING established for it a new genus (*Enhydra*); FISCHER in his synopsis endeavoured to bring this to the Greek (*Enydris*), which was also applied to it by LICHTENSTEIN.

The best generic descriptions of the Sea Otter that we have seen are those of the last named author, who has given two plates representing the skull and the teeth ; the latter however were deficient in number, owing to the fact of his specimen being a young animal with its dentition incomplete. In the Philosophical Transactions (1796, No. 17) we have a description of the anatomy of this animal by EVERARD HOME and ARCHIBALD MENZIES, which gives a tolerable idea of its structure.

There are only two authors, so far as we are aware, who have given reliable accounts of the habits of the Sea Otter—STELLER and COOK. The information published by the former is contained in Nov. Com. Acad. Petropolit., vol. ii. p. 267, ann. 1751 ; the latter gives an account of the animal in his Third Voyage, vol. ii. p. 295.

MUSTELA MARTES.—Linn.—Gmel.

PINE MARTEN.

PLATE CXXXVIII.—MALE and FEMALE. WINTER PELAGE.

M. Magnitudine Putorio visone major, flavida, hic illic nigrescens, capite pallidiore, gulâ flavescente, cauda longa, floccosa, acuta.

CHARACTERS.

Larger than the mink; general colour, yellowish, blended with blackish in parts; head, lighter; throat, yellow. Tail, long, bushy, and pointed.

SYNONYMES.

GENUS MUSTELA. Linn.
SUB-GENUS MUSTELA. Cuvier.
MUSTELA MARTES. Linn. Gmel., vol. i. p. 95.
PINE MARTEN. Pennant's Arctic Zoology, vol. i. p. 77.
MUSTELA MARTES. Sabine, Franklin's Journey, p. 651.
   "   "   Harlan, Fauna, p. 67.
   "   "   Godman, Nat. Hist., vol. i. p. 200.
   "   ZIBELLINA (?). Godman, Nat. Hist., vol. i. p. 208.
   "   MARTES. Rich., F. B. A., p. 51, summer specimen.
   "   HURO. F. Cuv.
   "   MARTES—AMERICAN SABLE. DeKay, Nat. Hist. State of New York, part i.
                     p. 32, pl. 19, fig. 2, skull.

DESCRIPTION.

Head, long and pointed; ears, broad and obtusely pointed; legs, rather long and tolerably stout; eyes, small and black; tail, bushy and cylindrical; toes, with long, slender, and compressed nails, nearly concealed by the hair. Hair, of two kinds—the outer long and rigid, the inner soft and somewhat woolly.

COLOUR.

This species varies a good deal in colour, so that it is difficult to find two specimens exactly alike; the under fur, however, does not differ as

Plate CXXXVIII

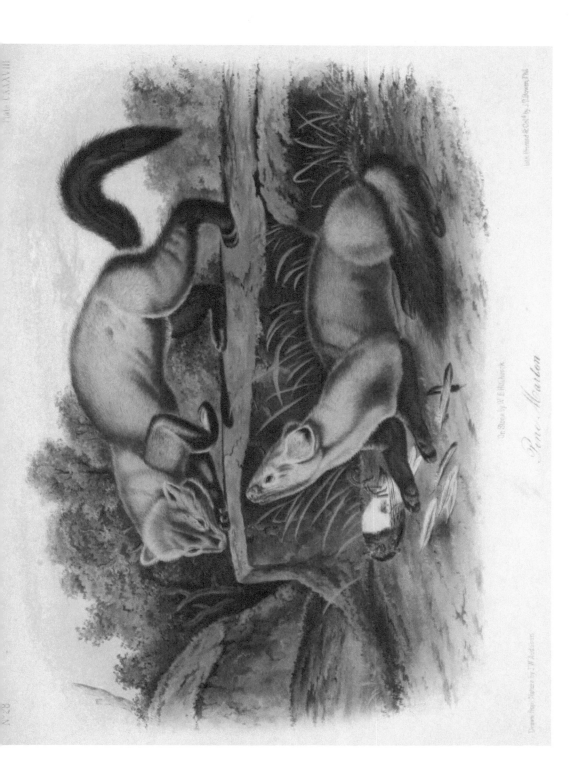

On Stone by W. E. Hitchcock

Pine Marton

Lith Printed & Col.d by J.T. Bowen, Phil.a

much in tint in different specimens as it does in fineness. Some individuals, particularly those captured in low latitudes, have much coarser fur than those from high northern regions or mountainous districts. The hair, which is about an inch and a quarter long, is of a pale dull grayish-brown from the roots outwards, dull yellowish-brown near the points, and is tipped with dark brown or black.

There is sometimes a considerable lustre in the fur of the Pine Marten ; the hair on the tail is longer, coarser, and darker than that on the body, and the coat is darkest in winter ; the yellowish-white markings on the throat vary in different individuals.

In the beginning of summer the dark-tipped hairs drop out, and the general colour of the fur is a pale orange brown, with little lustre ; the tips of the ears, at all times lighter than the rest of the fur, become very pale in summer. The feet are generally darker coloured than the hair of the body. The tip of the nose is flesh coloured ; eyes, black ; nails, light brown.

DIMENSIONS.

A winter-killed specimen, exceedingly poor.

	Foot.	Inches.
From point of nose to root of tail,	1	5
Length of tail (vertebræ),		7
" " (to end of hair),		10
" fore-leg to end of longest nail,		5¼
" hind-foot from heel to end of claws,		3
" ear on the outer surface,		1¼

We have measured larger specimens, 20, 21, and 22 inches from point of nose to root of tail.

HABITS.

Let us take a share of the cunning and sneaking character of the fox, as much of the wide-awake and cautious habits of the weasel, a similar proportion of the voracity (and a little of the fetid odour) of the mink, and add thereto some of the climbing propensities of the raccoon, and we have a tolerable idea of the attributes of the little prowler of which we have just given the description and dimensions. The Pine Marten, as may be inferred from this compound, is shy, cruel, cunning, and active, and partakes of the habits of the predacious animals above mentioned, with the exception that it is not known to approach the residences of man like the fox, weasel, or mink, but rather keeps in dense woods where it can prey upon birds, their eggs and young, squirrels, the white-footed and other mice, shrews, wood-rats, &c., together with beetles and other insects, larvæ of different species, toads, frogs, lizards, water reptiles, and fish. It is also an eater of some kinds of berries and nuts (as we are informed), and is said to be fond of honey like the bear.

It has been supposed that the name Pine Marten was given to this animal because it inhabits the pine forests of the northern parts of this continent, and shows a preference for those trees, in the lofty tops of which it frequently resides. The Pine Marten, however, is often called the American Sable or the Sable, and in fact is more generally known to the country people of our northern States, and also to the furriers, by the latter name than by any other.

Sprightly and agile in its movements, the Pine Marten commonly procures abundance of food. It is prolific, bringing forth from six to eight young at a time, so that notwithstanding the value of its fur and the consequent pursuit of it during the proper season, it is still by no means a scarce animal. We have had several specimens sent to us by friends residing in the State of New York and in the wilder portions of our Canada frontier, which were procured among the woody hills of those districts.

According to Dr. DeKay (New York Fauna, p. 33), this species is so active as to destroy great quantities of squirrels, the red squirrel (*Sciurus Hudsonius*) only escaping by its superior agility. Dr. Godman remarks that the "Pine Marten frequently has its den in the hollows of trees, but very commonly takes possession of the nest of some industrious squirrel, which it enlarges to suit its own convenience, after putting the builder to death."

Sir John Richardson says that "particular races of Martens, distinguished by the fineness and dark colour of their fur, appear to inhabit certain rocky districts." "A partridge's head, with the feathers, is the best bait for the log traps in which this animal is taken. It does not reject carrion, and often destroys the hoards of meat and fish laid up by the natives, when they have accidentally left a crevice by which it can enter. The Marten, when its retreat is cut off, shows its teeth, sets up its hair, arches its back, and makes a hissing noise like a cat. It will seize a dog by the nose and bite so hard, that unless the latter is accustomed to the combat, it suffers the little animal to escape."

The Indians sometimes eat the Pine Marten, but its flesh is rank and coarse. We have seen this species in confinement, when it appeared tolerably gentle, and had lost much of its snappish character.

The Pine Marten burrows in the ground at times, and the female brings forth her young in a fallen hollow log, a hole under rocks, or in a burrow, generally in April or May. These animals are chiefly caught with dead-falls baited with meat of any kind, birds, rabbits, squirrels, &c., and generally a hunter has many traps set, each of which he visits as often as once or twice a week. The Martens are sometimes devoured by larger animals after they have been caught. They are only trapped in the autumn and winter.

The fur of this species has been considered valuable, and when in fashion the skins were worth good prices. It is often palmed off on purchasers as fur of a more costly kind, and for this purpose is dyed any desired colour.

GEOGRAPHICAL DISTRIBUTION.

This species inhabits the wooded districts of the northern parts of America from the Atlantic to the Pacific in great numbers, and RICHARD-SON remarks that it is particularly abundant where the trees have been killed by fire but are still standing. HEARNE observed that it is very rare in the district lying north of Churchill river, and east of Great Slave lake. PENNANT states that on the Asiatic side of Behring's straits, twenty-five degrees of longitude in breadth are equally unfrequented by the Marten, and for the same reason—the absence of trees.

The limit of its northern range in America is, like that of the woods, about the 68th degree of latitude. It is found in the hilly and wooded parts of the northern Atlantic States. We have seen specimens obtained from near Albany and from the Catskill Mountains, and it is also found in the northern parts of Pennsylvania. Its southern limit is about lat. 40°.

We have sought for it in vain on the mountains of Virginia, where not-withstanding, we think a straggler will occasionally make its appearance. On the eastern continent it inhabits all the north of Europe and Asia.

GENERAL REMARKS.

Some American naturalists have expressed great doubts whether our American Marten is identical with that of the north of Europe, and have supposed that it might be designated under a separate specific name. We have not had an opportunity of comparing specimens from the two conti-nents with each other, as we could find no museum in which specimens from both continents were contained. We have, however, examined and taken descriptions of them separately, and have been able to detect so little difference that we cannot regard them even as varieties.

It has been frequently asserted by hunters, that the true Sable exists in America; thus far, however, no specimen of that animal has been identified as coming from this country. Those that were shown to us under the name of Sables by furriers, we ascertained to be fine skins of a very dark colour of our common Pine Marten.

Plate CXXXIX.

Drawn from Nature by J.W. Audubon.

On Stone by W.E. Hitchcock.

Lith. Printed & Col.d by J.T. Bowen, Phila.

Large tailed Spermophile.

SPERMOPHILUS MACROURUS.—Bennett.

LARGE-TAILED SPERMOPHILE.

PLATE CXXXIX.—Male.

S. Magnitudine Sciurum cinereum adequans, vellere crassiusculo, in dorso lateribusque cinereo nigroque vario, cauda corporis longitudine, mediocriter comosa.

CHARACTERS.

Size of the cat-squirrel (Sciurus cinereus) ; *fur, rather coarse ; body, mottled with black, and ashy white, forming irregular interrupted narrow transverse bars on the back and sides ; tail, as long as the body, and moderately bushy.*

SYNONYMES.

SPERMOPHILUS MACROURUS. Bennett, Proc. Zool. Soc., 1833, p. 41.
LONG-TAILED MARMOT. Zool. Soc. Catalogue, No. 456.
SCIURUS LUPTUS. Named in the Museum of the Jardin des Plantes, but not described.

DESCRIPTION.

This animal is shaped very much like a squirrel, although the ears are farther back in the head and the body is stouter than in that genus. Head, of moderate size, round, and elongated ; nose, somewhat pointed ; ears, large, broad, and ovate towards the points ; feet, stout ; nails, long, sharp, and considerably arched ; tail, rounded, possessing none of the distichous arrangement of the tails of squirrels ; tarsi, naked beneath ; fur, moderately long, and rather coarse and harsh to the touch.

COLOUR.

Hairs of the back, blackish-gray at the base, annulated with white, or brownish-white, towards the tips, which are black ; crown of the head, pure black ; muzzle, rufous brown above, whitish on the sides ; a narrow whitish space around the eyes ; on the lower part of the cheeks and on the throat the hairs are brownish-white ; cheeks, grizzled black and white ; ears, internally covered with short hairs and partly coloured on the inner surface with dusky and soiled yellow ; on the outside they are blackish-brown, becoming paler and grizzled towards the margins ; feet, whitish, finely freckled with dusky markings, their general hue pale ; tail, moderately bushy and sub-depressed ; the hairs are long, varying from one and three quarters to two inches in length ; they are of a brownish-white colour and are annulated by three broad black rings, the annulations nearest the apex of each hair considerably broader than the others.   Upper and lower incisors, pale yellow ; whiskers, black ; claws, brown.

In the specimen here described the whole crown of the head is black, but we are informed by our friend WATERHOUSE that an imperfect skin of a second specimen which exists in the museum of the Zoological Society of London has the crown of the head gray.

DIMENSIONS.

	Foot.	Inches.	Lines.
Length from point of nose to root of tail, - -	1	1	
" of tail (vertebræ), - - - - -		7	9
" " (including hair), - - - -		10	
" from nose to ear, - - - - -		2	5
Height of ear, - - - - - - -			6
Heel to end of claws, - - - - - -		2	5
Length of nail of middle hind-toe, - - - -			4½
" fore-foot and nails, - - - -		1	6
" nail of middle toe of fore-foot, - -			4½

HABITS.

Spermophilus Macrourus is an active and sprightly fellow, readily ascending trees on occasion, and feeding on nuts as well as seeds, roots, and grasses.

This species is in some districts rather numerous, and when in the rainy season some of the low grounds are submerged, takes to the trees, and sometimes curious fights occur between it and the wood-peckers. Five or six of the latter will on observing the Spermophile, unite against him, and cutting about in the air, peck at him as they dart swiftly around the persecuted animal, which is lucky if a hollow into which he can retreat be near, and frequently indeed the wood-peckers' holes are entered by him, but the angry and noisy birds still keep up their cries and fly with fury at the hole, and although they can no longer peck the animal they keep him in a state of siege for a considerable time.

The origin of this animosity may be the fact of the Spermophile (as well as many kinds of squirrels) sometimes turning out the wood-peckers from their nests, an injury which unites them against the wrong-doer. By what process the birds are influenced to attack when the animal is not in their nests, nor even on a tree upon which they have built (or dug, we should say), we know not, but that the birds comprehend that union is strength is quite evident, and the Spermophile knows it too, for he always instantly tries to escape and conceal himself as soon as the vociferous cries of the first bird that observes him are heard, and before its neighbours called thereby to the fight can reach the spot.

We have not been able to ascertain how many young this Spermophile produces at a birth, nor at what season they are brought forth. It is seen on the plains and in localities where no trees grow, in which places it burrows or runs into holes in the rocks.

From our present information we are inclined to think that this species is sometimes in company with *S. Douglasii* in California, or at least inhabits the same districts.

GEOGRAPHICAL DISTRIBUTION.

This Spermophile exists in some portions of that part of Mexico which were traversed by J. W. AUDUBON on his way towards California, and is also found in the last named State.

GENERAL REMARKS.

This species somewhat resembles *Spermophilus Douglasii*, but is a larger animal, the white patches over the shoulders moreover are wanting. The heel is hairy beneath, but the remaining part of the under surface of the foot is naked, whilst in *Spermophilus Douglasii* the whole foot is covered with hair beneath, up to the fleshy parts at the base of the toes.

PUTORIUS AGILIS.—Aud. and Bach.

PLATE CXL.—Male and Female. Winter Pelage.

P. Magnitudine intermedius P. pusillum inter et P. fuscum ; caudâ longâ, auriculis prominulis, æstate supra dilute fuscus, subtus albus, hyeme corpore toto caudaque niveis, cauda apice nigro.

CHARACTERS.

Intermediate in size between P. pusillus and P. fuscus ; tail, long ; ears, prominent. Colour, in summer, light brown above, white beneath ; in winter, body and tail, pure white, except the tip of the latter, which is broadly tipped with black.

DESCRIPTION.

This hitherto undescribed species is light, slender, and graceful, with well proportioned limbs, giving evidence of activity and sprightliness ; it may be termed a miniature of the ermine ; it stands proportionately higher on its legs, and although the smaller animal of the two, has the most prominent ears ; the hair is softer and shorter, both in summer and winter, than in either the ermine or Brown Weasel (*P. fuscus*) ; whiskers, numerous but rather short. Head, moderate ; skull, broad ; nose, short and rather pointed ; feet, small ; nails, partially concealed by the hair on the feet ; tail, long, covered with fur to within one and three quarters of an inch of the end, where it terminates in long straight smooth hairs.

COLOUR.

In summer : Head, ears, neck, outer surface of thighs, all the upper portions of the back, and the tail on both surfaces to near the tip, light brown, which is the colour of the hair from the roots to the tips ; end of the tail, black ; chin, throat, chest, belly, and inner side of thighs, white ; the brown colour extends far down on the sides and flanks, leaving a rather narrow stripe of white beneath, which is broadest on the neck ; the line of demarcation between the upper and under colours on the sides is dis-

Plate CXL.

Drawn from Nature by J W Audubon

On stone by W E Hitchcock

Lith⁴ Printed & Col⁴ by J.T. Bowen Philad⁴

Little Nimble Weasel.

tinctly but somewhat irregularly drawn. All the feet are brown ; whiskers and nails, dark brown ; teeth, white.

In winter : Pure white on the whole body, and for about three inches on the tail ; tip of the tail, black for an inch and three quarters ; tip of nose, flesh colour ; whiskers, mostly white, a few black.

DIMENSIONS.

	Inches.
Point of nose to root of tail, - - - - - - -	8½
Length of tail (vertebræ), - - - - - - -	3¾
" " (to end of hair), - - - - -	4¾
Point of nose to ear, - - - - - - -	1¼
Height of ear externally, - - - - - - -	¾

HABITS.

We preserved a specimen of this little animal during several months in the winter, forty years ago, in the northern part of New York ; it had been captured in a box trap, which was set near its hole in a pine forest, whither we had tracked it on the snow, believing from its small foot-prints that it was some unknown species of Rodentia. What was our surprise when on the following morning we discovered the eyes of this little marauder prying through the crevices of the trap. Supposing it to be a young ermine we preserved it through the winter, under the impression that it would become tame, and increasing in size, attain its full growth by the following spring ; we were, however, disappointed in our expectations ; it continued wild and cross, always printing on our gloves the form of the cutting edges of its teeth whenever we placed our hand within the box. It concealed itself in its nest, in a dark corner of the cage, during the whole day, and at night was constantly rattling and gnawing at the wires in the endeavour to effect its escape. We fed it on small birds, which it carried to its dark retreat and devoured greedily.

Having placed a common Weasel, twice the size of our animal, in the cage with it, the ermine immediately attacked our little fellow, which ensconced itself in a corner at the back of the cage, where with open mouth and angry eyes, uttering a hissing spitting or sputtering noise, he drew back his lips and showed his sharp teeth in defiance of his opponent.

To relieve him from a troublesome companion we removed the ermine. Towards spring we placed a Norway rat in his cage in order to test his courage. The rat and the Weasel retreated to opposite corners and eyed each other during the whole day ; on the following morning we found the rat had been killed; but the Weasel was so much wounded that he died before evening.

We have no other information in regard to the habits of this Weasel Its burrow, the entrance to which was very small, and without any hillock of earth at its borders, was situated in a high ridge of pine land.

We have no doubt that, like the ermine, in prowling about it finds its way into the retreats of the meadow-mouse, the little chipping squirrel, and other small animals, for although the rat above mentioned was too formidable an opponent, we are confident it could easily have mastered the little *Tamias Lysteri*.

GEOGRAPHICAL DISTRIBUTION.

We have only observed this Weasel in the northern part of the State of New York, but the specimens from which we drew our figures were procured by Mr. J. G. BELL in Rockland county in that State.

Plate CXLI

American Black Bear

URSUS AMERICANUS.—Pallas.

American Black Bear.

PLATE CXLI.—Male and Female.

U. Naso fere in eadem linea cum fronte, convexiore quam in U. feroce, plantis palmisque brevissimis, colore nigro vel fuscescente-nigro, lateribus rostri fulvis.

CHARACTERS.

Nose, nearly in a line with the forehead, more arched than in Ursus ferox : *palms and soles of the feet, very short ; colour, black, or brownish-black ; there is a yellowish patch on each side of the nose.*

SYNONYMES.

Black Bear. Pennant, Arctic Zoology, p. 57, and Introduction, p. 120.
  "   "   Pennant's History of Quadrupeds, vol. ii. p. 11.
  "   "   Warden's United States, vol. i. p. 195.
Ursus Americanus. Pallas, Spicil. Zool., vol. xiv. pp. 6–24.
  "   "   Harlan, Fauna, p. 51.
  "   "   Godman's Natural History, vol. i. p. 194.
  "   "   Rich., Fauna Boreali Americana, p. 14.
  "   "   DeKay, Nat. Hist. State of New York, p. 24, pl. 6. fig. 1.

DESCRIPTION.

The Black Bear is commonly smaller than the Grizzly Bear. Body and legs, thick and clumsy in appearance ; head, short, and broad where it joins the neck ; nose, slightly arched, and somewhat pointed ; eyes, small, and close to each other : ears, high, oval, and rounded at the tips ; palms and soles of the feet, short when compared with those of the Grizzly Bear : the hairs of the feet project slightly beyond the claws ; tail, very short : claws, short, blunt, and somewhat incurved ; fur, long, straight, shining, and rather soft.

COLOUR.

Cheeks, yellow, which colour extends from the tip of the nose on both sides of the mouth to near the eye ; in some individuals there is a small

spot of the same tint in front of the eye, and in others a white line commencing on the nose reaches to each side of the angle of the mouth ; in a
few specimens this white line continues over the cheek to a large white
space mixed with a slight fawn colour, covering the whole of the throat,
whence a narrow line of the fawn colour descends upon the breast. The
hairs on the whole body are in most specimens glossy black ; in some
we examined they were brown, while a few of the skins we have seen were
light brown or dingy yellow. From this last mentioned variety doubtless
originated the names Cinnamon Bear, Yellow Bear of Carolina, &c. The
outer edges of the ears are brownish-black ; eyes and nails, black.

DIMENSIONS.

A very large specimen.

	Feet.	Inches.
From nose to root of tail, · - · - -	6	5
Height to top of shoulder, - - · - -	3	1

A larger Bear than the above may sometimes be captured, but the
general size is considerably less.

HABITS.

The Black Bear, however clumsy in appearance, is active, vigilant, and
persevering, possesses great strength, courage, and address, and undergoes
with little injury the greatest fatigues and hardships in avoiding the pursuit
of the hunter. Like the deer it changes its haunts with the seasons, and
for the same reason, viz. the desire of obtaining suitable food, or of retiring
to the more inaccessible parts, where it can pass the time in security, unobserved by man, the most dangerous of its enemies.

During the spring months it searches for food in the low rich alluvial
lands that border the rivers, or by the margins of such inland lakes as, on
account of their small size, are called by us ponds. There it procures
abundance of succulent roots and tender juicy plants, upon which it chiefly
feeds at that season. During the summer heat, it enters the gloomy
swamps, passes much of its time in wallowing in the mud like a hog, and
contents itself with crayfish, roots, and nettles, now and then seizing on a
pig, or perhaps a sow, a calf, or even a full-grown cow. As soon as the
different kinds of berries which grow on the mountains begin to ripen,
the Bears betake themselves to the high grounds, followed by their cubs.

In retired parts of the country, where the plantations are large and the
population sparse, it pays visits to the corn-fields, which it ravages for a
while. After this, the various species of nuts, acorns, grapes, and other

forest fruits, that form what in the western States is called *mast*, attract its attention. The Bear is then seen rambling singly through the woods to gather this harvest, not forgetting, meanwhile, to rob every *bee-tree* it meets with, Bears being expert at this operation.

The Black Bear is a capital climber, and now and then *houses* itself in the hollow trunk of some large tree for weeks together during the winter, when it is said to live by sucking its paws.

At one season, the Bear may be seen examining the lower part of the trunk of a tree for several minutes with much attention, at the same time looking around, and snuffing the air. It then rises on its hind-legs. approaches the trunk, embraces it with the fore-legs, and scratches the bark with its teeth and claws for several minutes in continuance. Its jaws clash against each other until a mass of foam runs down on both sides of the mouth. After this it continues its rambles.

The female Black Bear generally brings forth two cubs at a time although, as we have heard, the number is sometimes three or four. The period of gestation is stated to be from six to seven weeks, but is mentioned as one hundred days by some authors. When born the young are exceedingly small, and if we may credit the accounts of hunters with whom we have conversed on the subject, are not larger than kittens. They are almost invariably brought forth in some well concealed den, or great hollow tree, and so cautious is the dam in selecting her place of accouchment, that it is extremely difficult to discover it, and consequently very rarely that either the female or her cubs are seen until the latter have obtained a much larger size than when born, are able to follow their dam, and can climb trees with facility.

Most writers on the habits of this animal have stated that the Black Bear does not eat animal food from choice, and never unless pressed by hunger. This we consider a great mistake, for in our experience we have found the reverse to be the case, and it is well known to our frontier farmers that this animal is a great destroyer of pigs, hogs, calves, and sheep, for the sake of which we have even known it to desert the pecan groves in Texas. At the same time, as will have been seen by our previous remarks, its principal food generally consists of berries, roots, and other vegetable substances. It is very fond also of fish, and during one of our expeditions to Maine and New Brunswick, we found the inhabitants residing near the coast unwilling to eat the flesh of the animal on account of its fishy taste. In our western forests, however, the Bear feeds on so many nuts and well tasted roots and berries, that its meat is considered a great delicacy, and in the city of New York we have generally found its market price three or four times more than the best beef per pound. The fore-paw of the Bear when cooked presents a striking resemblance to the hand of a child or young person, and we have known some individuals to be hoaxed by its being represented as such.

Perhaps the most acrid vegetable eaten by the Bear is the Indian turnip (*Arum triphyllum*), which is so pungent that we have seen people almost distracted by it, when they had inadvertently put a piece in their mouth.

The Black Bear is a remarkably swift runner when first alarmed, although it is generally " treed," that is, forced to ascend a tree, when pursued by dogs and hunters on horseback. We were, not very long since, when on an expedition in the mountains of Virginia, leisurely making our way along a road through the forest after a long hunt for deer and turkeys, with our gun thrown behind our shoulders and our arms resting on each end of it, when, although we had been assured there were no Bears in that neighbourhood, we suddenly perceived one above us on a little acclivity at one side of the road, where it was feeding, and nearly concealed by the bushes. The bank was only about fifteen feet high, and the Bear not more than twenty paces from us, so we instantly disengaged our gun, and cocking both barrels, expected to " fill our bag" at one shot, but at the instant and before we could fire, the Bear, with a celerity that astonished us, disappeared. We rushed up the bank and found the land on the top nearly level for a long distance before us, and neither very thickly wooded nor very bushy ; but no Bear was to be seen, although our eye could penetrate the woods for at least two hundred yards. After the first disappointing glance around, we thought Bruin might have mounted a tree, but such was not the case, as on looking everywhere nothing could be seen of his black body, and we were obliged to conclude that he had run out of sight in the brief space of time we occupied in ascending the little bank.

As we were once standing at the foot of a large sycamore tree on the borders of a long and deep pond, on the edge of which, in our rear, there was a thick and extensive " cane-brake," we heard a rushing roaring noise, as if some heavy animal was bearing down and passing rapidly through the canes, directly towards us. We were not kept long in suspense, for in an instant or two, a large Bear dashed out of the dense cane, and plunging into the pond without having even seen us, made off with considerable speed through the water towards the other shore. Having only bird-shot in our gun we did not think it worth while to call his attention to us by firing at him, but turned to the cane-brake, expecting to hear either dogs or men approaching shortly. No further noise could be heard, however, and the surrounding woods were as still as before this adventure. We supposed the Bear had been started at some distance, and that his pursuers not being able to follow him through the almost impenetrable canes, had given up the hunt.

Being one night sleeping in the house of a friend who was a Planter in the State of Louisiana, we were awakened by a servant bearing a light, who gave us a note, which he said his master had just received. We found it to be a communication from a neighbour, requesting our host and ourself to join him as soon as possible, and assist in killing some Bears at that moment engaged in destroying his corn. We were not long in dressing, and on entering the parlour, found our friend equipped. The overseer's horn was heard calling up the negroes. Some were already saddling our horses, whilst others were gathering all the cur-dogs of the plantation. All was bustle. Before half an hour had elapsed, four stout negro men, armed with axes and knives, and mounted on strong nags, were following us at a round gallop through the woods, as we made directly for the neighbour's plantation.

The night was none of the most favourable, a drizzling rain rendering the atmosphere thick and rather sultry ; but as we were well acquainted with the course, we soon reached the house, where the owner was waiting our arrival. There were now three of us armed with guns, half a dozen servants, and a good pack of dogs of all kinds. We jogged on towards the detached field in which the Bears were at work. The owner told us that for some days several of these animals had visited his corn, and that a negro who was sent every afternoon to see at what part of the enclosure they entered, had assured him there were at least five in the field that night. A plan of attack was formed : the bars at the usual entrance of the field were to be put down without noise ; the men and dogs were to divide, and afterwards proceed so as to surround the Bears, when, at the sounding of our horns, every one was to charge towards the centre of the field, and shout as loudly as possible, which it was judged would so intimidate the animals as to induce them to seek refuge upon the dead trees with which the field was still partially covered.

The plan succeeded : the horns sounded, the horses galloped forward, the men shouted, the dogs barked and howled. The shrieks of the negroes were enough to frighten a legion of bears, and by the time we reached the middle of the field we found that several had mounted the trees, and having lighted fires, we now saw them crouched at the junction of the larger branches with the trunks. Two were immediately shot down. They were cubs of no great size, and being already half dead, were quickly dispatched by the dogs.

We were anxious to procure as much sport as possible, and having observed one of the Bears, which from its size we conjectured to be the mother of the two cubs just killed, we ordered the negroes to cut down the tree on which it was perched, when it was intended the dogs should have a tug with it, while we should support them, and assist in preventing the Bear from escaping, by wounding it in one of the hind-legs. The surrounding woods now echoed to the blows of the axemen. The tree was large and tough, having been girded more than two years, and the operation of felling it seemed extremely tedious. However, at length it began to vibrate at each stroke; a few inches alone now supported it, and in a short time it came crashing to the ground.

The dogs rushed to the charge, and harassed the Bear on all sides, whilst we surrounded the poor animal. As its life depended upon its courage and strength, it exercised both in the most energetic manner. Now and then it seized a dog and killed him by a single stroke. At another time, a well administered blow of one of its fore-legs sent an assailant off, yelping so piteously that he might be looked upon as *hors du combat*. A cur had daringly ventured to seize the Bear by the snout, and was seen hanging to it, covered with blood, whilst several others scrambled over its back. Now and then the infuriated animal was seen to cast a revengeful glance at some of the party, and we had already determined to dispatch it, when, to our astonishment, it suddenly shook off all the dogs, and before we could fire, charged upon one of the negroes, who was mounted on a pied horse. The Bear seized the steed with teeth and claws, and clung to its breast. The terrified horse snorted and plunged. The rider, an athletic young man and a capital horseman, kept his seat, although only saddled on a sheep-skin tightly girthed, and requested his master not to fire at the Bear. Notwithstanding his coolness and courage, our anxiety for his safety was raised to the highest pitch, especially when in a moment we saw rider and horse come to the ground together; but we were instantly relieved on witnessing the masterly manner in which SCIPIO dispatched his adversary, by laying open his skull with a single well directed blow of his axe, when a deep growl announced the death of the Bear.

In our country no animal, perhaps, has been more frequently the theme of adventure or anecdote than the Bear, and in some of our southwestern States it is not uncommon to while away the winter evenings with Bear stories that are not only interesting on account of the traits of the habits of the animal with which they are interspersed, but from the insight they afford the listener into the characteristics of the bold and hardy huntsmen of those parts.

In the State of Maine the lumbermen (wood-cutters) and the farmers set guns to kill this animal, which are arranged in this way: A funnel-shaped space about five feet long is formed by driving strong sticks into the ground in two converging lines, leaving both the ends open, the narrow end being wide enough to admit the muzzle of an old musket, and the other extremity so broad as to allow the head and shoulders of the Bear to enter. The gun is then loaded and fastened securely so as to deliver its charge facing the wide end of the enclosure. A round and smooth stick is now placed behind the stock of the gun, and a cord leading from the trigger passed around it, the other end of which, with a piece of meat or a bird tied to it (an owl is a favourite bait), is stretched in front of the gun, so far that the Bear can reach the bait with his paw. Upon his pulling the meat towards him, the string draws the trigger and the animal is instantly killed.

On the coast of Labrador we observed the Black Bear catching fish with great dexterity, and the food of these animals in that region consisted altogether of the fishes they seized in the edge of the water inside the surf. Like the Polar Bear, the present species swims with ease and rapidity, and it is a difficult matter to catch a full grown Bear with a skiff, and a dangerous adventure to attempt its capture in a canoe, which it could easily upset.

We were once enjoying a fine autumnal afternoon on the shores of the beautiful Ohio, with two acquaintances who had accompanied us in quest of some swallows that had built in a high sandy bank, when we observed three hunters about the middle of the river in a skiff, vigorously rowing, the steersman paddling too, with all his strength, in pursuit of a Bear which, about one hundred and fifty yards ahead of them, was cleaving the water and leaving a widening wake behind him on its unrippled surface as he made for the shore, directly opposite to us. We all rushed down to the water at this sight, and launching a skiff we then kept for fishing, hastily put off to intercept the animal, which we hoped to assist in capturing. Both boats were soon nearing the Bear, and we, standing in the bow of our skiff, commenced the attack by discharging a pistol at his head. At this he raised one paw, brushed it across his forehead, and then seemed to redouble his efforts. Repeated shots from both boats were now fired at him, and we ran alongside, thinking to haul his carcase triumphantly on board; but suddenly, to our dismay, he laid both paws on the gunwale of the skiff, and his great weight brought the side for an instant under water, so that we expected the boat would fill and sink. There was no time to be lost: we all threw our weight on to the other side, to counterpoise that of the animal, and commenced a pell-mell battery on him with the oars and a boat-hook; the men in the other boat also attacked him, and driving the bow of their skiff close to his head, one of them laid his skull open with an axe, which killed him instanter.   We jointly hurraed, and tying a rope round his neck, towed him ashore behind our boats.

The Black Bear is very tenacious of life, and like its relative, the Grizzly Bear, is dangerous when irritated or wounded.   It makes large beds of leaves and weeds or grasses, in the fissures of rocks, or sleeps in hollow logs, when no convenient den can be found in its neighbourhood ; it also makes lairs in the thick cane-brakes and deep swamps, and covers itself with a heap of leaves and twigs, like a wild sow when about to litter.

The skin of the Black Bear is an excellent material for sleigh-robes, hammer-cloths, caps, &c., and makes a comfortable bed for the backwoods-man or Indian ; and the grease procured from this species is invaluable to the hair-dresser, being equal if not superior to

"Thine incomparable oil Macassar!"

which we (albeit unacquainted with the mode of preparing it) presume to be a compound much less expensive to the manufacturer than would be the " genuine real Bear's grease"—not of the shops, but of the prairies and western woods.

The Black Bear is rather docile when in confinement, and a " pet" Bear is occasionally seen in various parts of the country.   In our large cities, however, where civilization (?) is thought to have made the greatest advances, this animal is used to amuse the gentlemen of the fancy, by putting its strength and " pluck" to the test, in combat with bull-dogs or mastiffs.   When the Bear has not been so closely imprisoned as to partially destroy his activity, these encounters generally end with the killing of one or more dogs ; but occasionally the dogs overpower him, and he is rescued for the time by his friends, to " fight (again) some other day."

We are happy to say, however, that Bear-baiting and bull-baiting have not been as yet fully naturalized amongst us, and are only popular with those who, perhaps, in addition to the natural desire for excitement, have the hope and intention of winning money, to draw them to such cruel and useless exhibitions.

Among the many Bear stories that have been published in the newspa-pers, and which, whether true or invented, are generally interesting, the following is one of the latest, the substance of which we will give, as nearly as we can recollect it :

A young man in the State of Maine, whilst at work in a field, accompa-nied only by a small boy, was attacked by a Bear which suddenly approached from the edge of the forest, and quite unexpectedly fell upon him with great fury.   Almost at the first onset the brute overthrew the young farmer, who fell to the ground on his back, with the Bear clutching him, and biting his arm severely. Nothing but the utmost presence of mind could have saved the young man, as he was unarmed with the exception of a knife, which he could not get out of his pocket owing to the position in which he had fallen. Perceiving that his chance of escape was desperate, he rammed his hand and arm so far down the throat of the Bear as to produce the effect of partial strangulation, and whilst the beast became faint from consequent loss of breath, called to the boy to come and hand him the knife. The latter bravely came to the rescue, got the knife, opened it, and gave it to him, when he succeeded in cutting the Bear's throat, and with the exception of a few severe bites, and some lacerations from the claws of the animal, was not very much injured. The Bear was carried next day in triumph to a neighbouring village, and weighed over four hundred pounds.

Such assaults are, however, exceedingly rare, and it is seldom that even a wounded Bear attacks man.

Captain J. P. McCown has furnished us with the following remarks: " In the mountains of Tennessee the Bear lives principally upon mast and fruits. It is also fond of a bee-tree, and is often found seeking even a wasp's or yellow-jacket's nest. In the autumn the Bear is hunted when ' lopping' for chesnuts. Lopping consists in breaking off the branches by the Bear to procure the mast before it falls. When pursued by the dogs the Bear sometimes backs up against a tree, when it exhibits decided skill as a boxer, all the time looking exceedingly good-natured ; but woe to the poor dog that ventures within its reach!

" The dogs generally employed for pursuing the Bear are curs and fice, as dogs of courage are usually killed or badly injured, while the cur will attack the Bear behind, and run when he turns upon him. No number of dogs can kill a Bear unless assisted by man.

" In 1841, the soldiers of my regiment had a pet he-Bear (castrated) that was exceedingly gentle and playful with the men. It becoming necessary to sell or kill it, one of the soldiers led it down the streets of Buffalo and exposed it for sale. Of course it attracted a large crowd, and was bid for on all sides on account of its gentleness. But unfortunately Bruin was carried near a hogshead of sugar, and not disposed to lose so tempting a repast, quietly upset it, knocking out the head, and commenced helping himself in spite of the soldier's efforts to prevent the depredation. The owner of the sugar rushed out and kicked the Bear, which, not liking such treatment, gave in return for the assault made upon him, a blow that sent his assailant far into the street, to the terror of the crowd, which scattered, leaving him to satisfy his appetite for sugar unmolested."

The number of Black Bears is gradually decreasing in the more settled parts of the " back woods," but in some portions of Carolina and Georgia, where the vast swamps prevent any attempt to settle or cultivate the land, they have within a few years been on the increase, and have become destroyers of the young stock of the Planter (which generally range through the woods) to a considerable extent.

Sir JOHN RICHARDSON says that when resident in the fur countries this Bear almost invariably hibernates, and that about one thousand skins are annually procured by the Hudson's Bay Company from those that are destroyed in their winter retreats. " It generally selects à spot for its den under a fallen tree, and having scratched away a portion of the soil, retires to it at the commencement of a snow-storm, when the snow soon furnishes it with a close, warm covering. Its breath makes a small opening in the den, and the quantity of hoar-frost which occasionally gathers round the aperture serves to betray its retreat to the hunter."

The Black Bear is somewhat migratory, and in hard winters is found to move southwardly in considerable numbers, although not in company. They couple in September or October, after which the females retire to their dens before the setting in of very cold weather.

It is said that the males do not so soon resort to winter quarters as the females, and require some time after the love season to recover their lost fat. The females bring forth about the beginning of January.

The Indian tribes have many superstitions concerning the Bear, and it is with some of them necessary to go through divers ceremonies before proceeding to hunt the animal.

GEOGRAPHICAL DISTRIBUTION.

The Black Bear has been found throughout North America in every wooded district from the north through all the States to Mexico, but has not hitherto been discovered in California, where it appears to be replaced by the Grizzly Bear (*Ursus ferox*).

GENERAL REMARKS.

This species was in the early stages of natural history regarded as identical with the Black Bear of Europe. PALLAS first described it as a distinct animal, since which its specific name has remained undisturbed ; its varieties have however produced much speculation, and it has frequently been supposed, and not without some reason, that the Brown Bear of our western country was a species differing from the Black Bear.

In order to arrive at a correct conclusion on this subject we must be guided less by colour than by the form and structure of the animal and its length of heel and claws ; it is evident that the size can afford us no clue whereby to designate the species, inasmuch as some individuals may be found that are nearly double the dimensions of others.

PSEUDOSTOMA BOREALIS.—Rich. MSS.

The Camas Rat.

PLATE CXLII. Male, Female, and Young.

P. Ex cinereo fulvus, cauda longa pilosa ; P. bursario minor, et gracilior, dentibus unguibusque minoribus.

CHARACTERS.

Smaller and of more delicate form than Pseudostoma bursarius, *and teeth and claws much smaller. Tail, long, and clothed with hair. Colour, pale yellowish-gray.*

SYNONYMES.

Geomys Borealis. Rich., MSS.
Pseudostoma Borealis. Bach., Jour. Acad. Nat. Sciences Philadelphia, vol. vii
part 1, p. 103.
Geomys Townsendii. Rich., MSS.

DESCRIPTION.

Head, of moderate size ; ears, consisting of a small round opening mar-gined by an elevated ridge, the highest portion of which is the posterior part, and is about one line in height. The ears not hidden by the fur, but distinctly visible. Body, moderately thick ; claws of the fore-feet, slender and rather long ; incisors, rather long (but not large for the genus) ; the upper ones have each a slight longitudinal groove situated close to the inner margin. Tip of nose, naked ; feet, bare beneath ; inner toe of fore-feet, rather short, outer next in length, middle longest, and the toes on either side of the central one about equal ; there is a long brush of stiff white hairs on the inner side of the inner toe. On the hind-feet the central toe is longest, outer toes equal and short. Tail, hairy.

COLOUR.

General colour, pale gray, the upper parts more or less washed with yellow ; inside of pouches, under surface of body, feet, and tail, white.

Plate CXLII

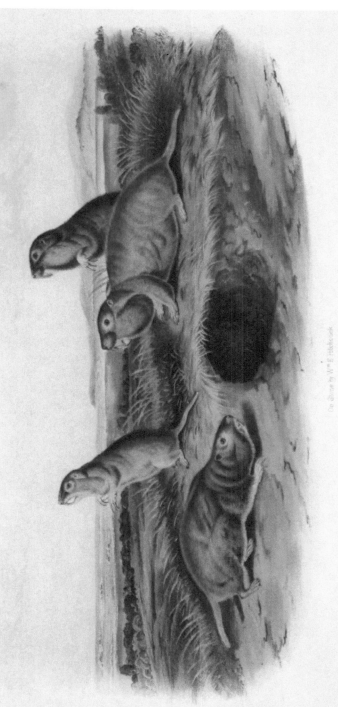

Drawn from Nature by J. W. Audubon.

On Stone by Wm E. Hitchcock

The Camas Rat

Lith. Printed & Col by J. T. Bowen, Phil.

Hairs of the body, dark slate colour at the roots. There is a dusky spot behind the ears ; incisors, yellow ; claws, white ; tail, above, grayish, tinged with yellow.

DIMENSIONS.

	Inches.	Lines.
From nose to root of tail, - - - - -	7	6
Tail, - - - - - - - -	2	
Tarsus and claws, - - - - -	1	1½
Central claw of fore-foot, - - - - -		5
Nose to ear, - - - - - -	1	6½

The above description was made from three specimens of this pouched Sand-rat, obtained by the late Mr. TOWNSEND, on the Columbia river, two of which appeared to be in summer pelage, and the third in its autumnal coat.

Description of another specimen sent by Mr. TOWNSEND, marked in RICHARDSON's MSS. as *Geomys Townsendii* :

Form and size of the animal, nearly the same as in the specimens just described, with the exception of the tail, which is considerably longer. General colour, very pale gray above, with a faint yellowish wash ; end of nose, dusky gray ; under parts, grayish-white ; chin, pure white ; tail and feet, white, the former grayish above. Hairs of the back, very pale gray at the roots, pale yellow near the tips, the extreme points cinereous. Teeth, yellowish-white ; upper incisors, with a faint groove near the internal margin. Claws and fore-feet, moderate white.

DIMENSIONS.

	Inches.	Lines.
From nose to tail, - - - - -	7	6
Tail, - - - - - - - -	2	9
Tarsus, - - - - - - -	1	3½
Central claw of fore-foot, - - - -		5
Nose to ear, - - - - - -	1	5

HABITS.

The Camas Rat derives its name, according to RICHARDSON, from its fondness for the bulbous root of the quamash or camas plant (*Scilla esculenta*).

Like all the pouched Rats of America, it feeds upon nuts, roots, seeds,

and grasses, and makes burrows, extending long distances, but not very far
beneath the surface of the ground, throwing up mole-hills in places as it
comes to the surface. These animals are generally found to be in a certain
degree gregarious, or at least a good many of them inhabiting the same
locality, and more or less associated together; and are said to be very
common on the plains of the Multnomah river.

Mr. DOUGLAS informed Sir JOHN RICHARDSON that they may be easily
snared in the summer.

We believe that some of the Indians of those parts of Oregon in which
this burrowing Rat exists eat them, but have no information concerning the
peculiarities they exhibit, the number of young they produce at a time, or
the depredations they commit on the fields and gardens of the settlers.

In the Fauna Boreali Americana (p. 206), this pouched Rat (if we are
not mistaken), is given as *Diplostoma bulbivorum*—Camas Rat—and under
the impression that that name applies to our present animal, we have
made the above remarks in relation to it.

GEOGRAPHICAL DISTRIBUTION.

Specimens were obtained both by DOUGLAS and DRUMMOND, about the
same period of time, in the vicinity of the Columbia river in Oregon.

GENERAL REMARKS.

On a visit to Europe we carried with us three specimens of pouched
Sand-Rats, which we regarded as belonging to the same species, but being
male, female, and young. Our object was to compare them with specimens
taken from this country at the north and west by RICHARDSON, DOUGLAS,
DRUMMOND, and other naturalists. RICHARDSON kindly showed us a
specimen brought from the Columbia river by DOUGLAS, which, as we
thought, appeared to be of the same species as our own. As he was then
preparing a monograph of this perplexing genus, we requested him to
describe the species, and add it to his monograph; he consequently gave
it the above name. He however called another specimen which we had
carried with us, *Geomys Townsendii*. We think his monograph was never
published.

We have united what he considered two species—*Geomys Borealis* and
G. Townsendii—into one, having added the latter as a synonyme; and we
have rejected *Diplostoma* as a genus, not only because we conceive the
characters on which it is founded to be the result of an unnatural disposition of the pouches in the dried skins, but for the reason mentioned above

viz., that we consider the so-called *Diplostoma bulbivorum* to be identical with the animal we have just described as *Pseudostoma borealis*, although the description given by RICHARDSON has apparently no reference to the latter, but on the contrary describes his *Diplostoma* as having the true mouth *vertical* (?). He says: "The lips, which in fact are right and left, and not upper and under," &c. Besides, in the beginning of his article he mentions that the skull is wanting. We think we may therefore reasonably presume, that although the skin had been so twisted and disfigured by putting it into an unnatural form that the appellation which Mr. DOUGLAS gave it, as "the animal known on the banks of the Columbia by the name of the *Camas Rat*," did not seem to apply to it, we shall be right in rejecting both the generic and specific names given by our friend Sir JOHN RICHARDSON to so very imperfect a specimen, and in believing that the skin was in reality (although much injured and distorted) nothing but the Camas Rat, as DOUGLAS called it.

PTEROMYS SABRINUS.—Pennant

SEVERN RIVER FLYING-SQUIRREL.

PLATE CXLIII.—Fig. 1.

P. Magnitudine P. volucellum tertia parte excedens ; caudâ corpore curtiore, patagio lumbari pone carpum in lobum rotundatum excurrente, colore flavescente-cano obscuriore inumbrato.

CHARACTERS.

One third larger than P. volucella ; *tail, shorter than the body ; flying membrane having a small rounded projection behind the wrist. Colour, dull yellow gray, irregularly marked with darker.*

SYNONYMES.

GREATER FLYING-SQUIRREL. Forster, Philos. Trans., vol. lxii, p. 379.
SEVERN RIVER FLYING-SQUIRREL. Pennant, Hist. Quad., vol. ii. p. 153.
    "   "   " Arctic Zoology, vol. i. p. 122.
SCIURUS HUDSONIUS. Gmel., Syst., vol. i. p. 153.
  " SABRINUS. Shaw, Zool., vol. ii., part 1, p. 157.
PTEROMYS SABRINUS. Rich., Zool. Jour., No. 12, p. 519.
  "   "   " F. B. A., p. 193.

DESCRIPTION.

Head, short and somewhat rounded ; nose, short and obtuse ; eyes, large ; flying membrane, extending from the wrist to the middle of the hind-leg, nearly straight, having only a slight rounded projection close to the wrist ; tail, depressed, slightly convex on its upper surface, but quite flat, or even somewhat concave, beneath ; it is broadest about an inch from the body, and then tapers gradually but very slightly towards the extremity, which is rounded ; the flattened form of the tail, and its distichous arrangement, is given to it in consequence of the fur on its sides being much longer than that on its upper surface ; the extremities are small ; the fore-legs connected with the flying membrane down to the wrist ; the feet are hairy both above and below. There are four short toes on the fore-feet, and the claws are small, compressed, curved, and sharp pointed ;

under their roots there is a compressed callous space, projecting from the end of each toe, and there is a callosity in place of a thumb, armed with a very minute nail.

There are five hind toes; the claws resemble those of the fore feet, and are almost concealed by the hair of the toes; the soles are covered with a dense brush, like the feet of a rabbit or hare. The fur is soft, long, and silky on all parts.

COLOUR.

Incisors, deep orange; whiskers, black; a dark gray marking around the eye. The hairs on the upper surface of the head and body are of a deep blackish-gray colour from the roots to near the tips, which are pale reddish-brown, but distinctly presented only when the fur lies smoothly; on the flying membrane the colour is a shade darker in consequence of the under colour not being concealed by the lighter colour of the tips; the outer surfaces of the feet are pale bluish-gray; the margins of the mouth, sides of the nose, cheeks, and whole ventral aspect of the body, white, with a tinge of buff under the belly, and particularly under the flying membrane. Tail, nearly the colour of the back, with an intermixture however of black hairs; beneath, it is buff; hair on the soles, yellowish-white.

DIMENSIONS.

	Inches.	Lines.
Length of head and body, - - - - -	8	
Tail, including fur, - - - - - -	5	9
Height of ear, - - - - - - -		5½
Heel to end of claw, - - - - - -	1	5½
Longest hind-toe and nail, - - - - -		4¾
Fore-toe and nail, - - - - - -		5

HABITS.

We found this interesting Flying-Squirrel in abundance at Quebec, and many of them were offered for sale in the markets of that city during our sojourn there. It appears indeed to take the place of the common small Flying-Squirrel of the United States (**P. volucella**) in Lower Canada, where we did not observe the latter east of Montreal.

We heard that one of these pretty animals was caught alive by a soldier who saw it on the plains of Abraham, and ran it down.

A brood of young of this species, along with the mother was kept in confinement by an acquaintance of ours, for about four months, and the little ones, five in number, were suckled in the following manner : the younglings stood on the ground floor of the cage, whilst the mother hung her body downwards, and secured herself from falling by clinging to the perch immediately above her head by her fore-feet. This was observed every day, and some days as frequently as eight or ten times.

This brood was procured as follows : a piece of partially cleared wood having been set on fire, the labourers saw the Flying-Squirrel start from a hollow stump with a young one in her mouth, and watched the place where she deposited it, in another stump at a little distance. The mother returned to her nest, and took away another and another in succession, until all were removed, when the wood-cutters went to the abode now occupied by the affectionate animal, and caught *her* already *singed* by the fire, and her five young unscathed.

After some time a pair of the young were given away to a friend. The three remaining ones, as well as the mother, were killed in the following manner :

The cage containing them was hung near the window, and one night during the darkness, a rat, or rats (*mus decumanus*), caught hold of the three young through the bars, and ate off all their flesh, leaving the skins almost entire, and the heads remaining inside the bars. The mother had had her thigh broken and her flesh eaten from the bone, and yet this good parent was so affectionately attached to her brood that when she was found in this pitiable condition in the morning, she was clinging to her offspring, and trying to nurse them as if they had still been alive.

This species is said to bear a considerable resemblance to the European Flying-Squirrel. It was first described by FORSTER, who not having distinguished it from the European animal, PENNANT stands as its discoverer.

We did not observe any of these Flying-Squirrels on the borders of the Yellow Stone or Upper Missouri, and have no further information as to their habits.

In our first volume (pp. 134, 135), we mentioned that Sir JOHN RICH-ARDSON speaks of a Flying-Squirrel which he considered a variety of *P. sabrinus*, and called *var. B. alpinus*. We then remarked that we hoped to be able to identify that variety when presenting an account of the habits of *P. sabrinus*, and in our next article shall have the pleasure of doing so, having named it *P. alpinus*.

GEOGRAPHICAL DISTRIBUTION.

The northern range of this species is about latitude 52° ; it has been

Fig 1

Fig 2

Fig 1   Severn River Flying Squirrel

Fig 2   Rocky Mountain Flying Squirrel

Drawn from Nature by J.W. Audubon   Lith. Printed & Col.d by J.T. Bowen, Phil.a

captured on the shores of Lake Huron, and at the bottom of James Bay, at Moose Factory. We obtained specimens in the neighbourhood of Quebec, where in the autumn they were exceedingly abundant.

We have not a doubt it is found in the United States south of the river St. Lawrence, but at present have no evidence to that effect. It does not appear to exist on either slope of the Rocky Mountains, nor have we in fact been able to find any of our smaller Rodentia of the Atlantic States in those regions.

GENERAL REMARKS.

As long as only two species of Flying-Squirrel were known in North America—the present species (*P. sabrinus*) and the little *P. volucella*—there was no difficulty in deciding on the species, but since others have been discovered in the far west, the task of separating and defining them has become very perplexing. We will however endeavour, in our next article, in which we shall describe *P. alpinus*, to point out those characters which may enable naturalists to distinguish the closely allied species.

PTEROMYS ALPINUS.—Aud. and Bach.

Rocky Mountain Flying-Squirrel.

PLATE CXLIII.—Fig. 2.

P. Magnitudine P. sabrino major, caudâ planâ, latâ, corpore longiore, patagio lumbari angusto, margine recta.

CHARACTERS.

Larger than Pteromys sabrinus ; *tail, flat and broad, longer than the body ; flying membrane, short and with a straight border.*

SYNONYME.

Pteromys Sabrinus (Var. Alpinus). Rich., F. B. A., p. 195, pl. 18.

DESCRIPTION.

Head longer and body stouter than in *P. sabrinus ;* the tail is also longer, much broader, more densely clothed with hair, and has a flatter and more elliptical form ; the flying membrane is much smaller than in *P. sabrinus*, and the border is straight ; the ears are thin and membranous, have a little fur at the base on the upper surface, and are thinly covered on both sides with short adpressed hairs ; their form is semi-oval with rounded tips ; the tail is flat, oblong, and oval in form ; the extremities are rather stout, more especially the hind-feet ; the soles, palms, and under surfaces of the toes are well covered with fur, except a small callous eminence at the end of each toe. There are five eminences on the palm, of which the two posterior ones are the largest ; and four on the soles, situated at the root of the toes. There is a brush of soft fur near the outer edges of the soles ; the fur is dense, very long, and has a woolly appearance ; the longest hair on the back is fully an inch in length.

COLOUR.

Head, nose, and cheeks, light grayish, with a slight wash of yellow ; surface of the fur on the back, yellowish-brown, without any tendency to the more red hue of the back in *P. sabrinus.*

The fur of the throat and belly is a grayish-white, without any tinge of buff colour ; tail, blackish-brown above, a little paler beneath.

DIMENSIONS.

	Inches.
From point of nose to root of tail, - - - - -	8¼
Tail (vertebræ), - - - - - - -	5
" (including fur), - - - - - - -	6¼
Heel to longest middle toe, - - - - - -	1½
Height of ear posteriorly, - - - - - -	⅜
Breadth between the outer edges of the flying membrane,	4¾

RICHARDSON states that there is a specimen in the Hudson's Bay Museum, which measures nine inches from the point of the nose to the root of the tail.

HABITS.

We have learned little of the habits of this animal. DRUMMOND, who obtained it on the Rocky Mountains, states that it lives in pine forests, seldom venturing from its retreats except during the night.

From its heavy structure, and the shortness of the bony process that supports the flying membrane, we are led to infer that it is less capable of supporting itself in sailing from one tree to another, than the other species of this genus.

GEOGRAPHICAL DISTRIBUTION.

Both the specimens of DRUMMOND and TOWNSEND were obtained in crossing the Rocky Mountains on the usual route to the Columbia river. We have no doubt this species will be found on the western side of the Rocky Mountains, from the Russian settlements through Oregon to California.

GENERAL REMARKS.

RICHARDSON regarded this species as a variety of *Pteromys sabrinus* (see our first volume, p. 134), and adopted for it the name *alpinus*, not to designate a species but a variety. We, on the other hand, consider it a true species, and have applied to it the name of *P. alpinus*, quoting RICHARDSON'S *var. alpinus* as a synonyme.

On comparing the specimen from which our drawing was made, with *P.*

sabrinus from Quebec, the following appeared to be the points of difference : *alpinus* is considerably the larger animal, and although the legs appear somewhat shorter, they are stouter ; the fur is more dense and longer, having quite a woolly appearance ; the ears are shorter than in *P. sabrinus*, and are broader and more rounded. They may also be distinguished by the colour of their fur from each other, that of *alpinus* on the under surface being pure white from the roots, while the fur of *P. sabrinus* is tinged with yellowish. The most striking difference, however, is the extreme shortness of the bony process which supports the flying membrane at the fore-leg.

ARVICOLA TOWNSENDII.—Bach

TOWNSEND'S ARVICOLA.

PLATE CXLIV. Fig. 1.—MALE.

A. Mure decumano duplo minor, auriculis erectis, vellere prominulis, colore in dorso plumbeo ad rufum vergente in capite colloque.

CHARACTERS.

Half the size of the Norway rat ; ears, upright, and visible beyond the fur ; Plumbeous on the back, inclining to rufous on the head and neck.

SYNONYME.

ARVICOLA TOWNSENDII. Bachman, Jour. Acad. Nat. Sciences, vol. viii., part 1, p. 60.

DESCRIPTION.

Body, cylindrical ; head, rather small ; whiskers, long, reaching beyond the ears ; eyes, small ; teeth, large ; ears, large, broad, erect, extending considerably above the fur ; feet, of moderate size ; toes, like the rest of this genus ; thumb, protected by a rather short acute nail ; tail, scaly, sparingly covered with soft hair, a few hairs at its extremity ; feet, clothed to the nails with short brown adpressed hairs ; fur, on the back, about three lines long, much shorter beneath.

COLOUR.

Whiskers, white and black ; teeth, yellow ; fur on the upper part of the body, lead colour from the roots to near the tips, which present a mixture of white and black points, from which results a general plumbeous colour ; under surface, grayish-ash ; neck, sides of face, nose, and an obscure line above the eye, ashy-brown ; tail, brownish, with a few white hairs at the tip ; feet, yellowish-brown ; claws, brown.

DIMENSIONS.

	Inches.	Lines.
Length of head and body, - - - - - -	6	
Tail, - - - - - - - - -	2	6

VOL. III.—27

	Inches.	Lines.
Fore-feet to point of nails, - - - - - - -		9
Heel to point of nail, - - - - - - -	1	
Breadth of ear, - - - - - - - -		5

HABITS.

The late Mr. Townsend, who captured this animal under an old log on the banks of the Columbia river, gave us no account of its habits. We should judge from its form, its conspicuous ears, and its general resemblance to the cotton rat of Carolina (*Sigmodon hispidum*), that it possesses many of its characteristics. It was found in the woods, but we imagine that it exists on the edges of the open country skirting the forests, feeding on roots, grasses, and seeds, nestling under logs and brushwood, and having, like the rest of the genus, four or five young at a birth.

GEOGRAPHICAL DISTRIBUTION.

The specimen here described was obtained on the 21st of July, 1835, by Mr. Townsend, on the shores of the Columbia river. It no doubt is widely distributed on the western side of the Rocky Mountains, and replaces the Wilson's Meadow-Mouse of our northern Atlantic States.

GENERAL REMARKS.

We find it exceedingly difficult to ascertain characters to designate the various species of Arvicolæ in our country; they resemble each other in many particulars, and especially in colour. We can however find no description which answers to this species in Richardson or any other author.

ARVICOLA NASUTA.—Aud. and Bach.

SHARP-NOSED ARVICOLA.

PLATE CXLIV. Fig. 2.—Male.

A. A. Pennsylvanica longior, caudâ, capite, breviore ; pedibus, tenuibus ; calce brevissima ; corpore supra, ferrugineo-fusco ; subtus ex cinereo et flavo variegato.

CHARACTERS.

Larger than Arvicola Pennsylvanica ; *tail shorter than the head ; legs small and slender ; heel very short ; body above, dark rusty brown, soiled yellowish-gray beneath.*

DESCRIPTION.

The head of this species is rather longer and the nose sharper than in the Arvicolæ in general ; the lower incisors are long and very much curved ; the body is less cylindrical than that of Wilson's Meadow-Mouse ; ears, circular, sparingly hairy within, and well covered with fur exteriorly ; whiskers, shorter than the head ; tail, thinly covered with hair.

The legs are rather slender, and are covered with short hairs ; the forefeet have naked palms ; claws, small ; tarsus, more than one third shorter than that of the much smaller *Arvicola Pennsylvanica ;* the fur on the back is also shorter.

COLOUR.

Incisors, yellowish-white ; fur on the back, from the roots to near the tips, grayish-black ; the tips are yellowish-brown and black, giving it a rusty brown appearance ; the legs and tail are light brown ; the chin, soiled white ; the fur on the under surface of the body is dark cinereous from the roots to near the tips, where it is light coloured.

DIMENSIONS.

	Inches.	Lines.
Length of head and body, - - - - - -	5	9
Head, - - - - - - - - -	1	10

	Inches.	Lines.
Tail,	1	6
Heel to point of nail,		6

For the sake of convenient comparison we give the dimensions of the largest of six specimens of Arvicola Pennsylvanica :

	Inches.	Lines.
Length of head and body,	4	2
Head,	1	4
Tail,	1	6
Heel to point of longest nail,		11

HABITS.

We have found this species breeding in the vicinity of Wilson's Meadow-Mouse, although never nearer than a few hundred yards from the latter, and we have sometimes observed their nests in summer on large hillocks of sedge-grass (*carex*) growing in marshy localities, and surrounded by water ; they do not occupy these exposed situations, however, in winter, but are found on more elevated knolls, under the roots of old trees or shrubs. They produce four or five young at a birth, and certainly breed twice, if not oftener, during the season.

GEOGRAPHICAL DISTRIBUTION.

The specimen which we have described was obtained by Dr. BREWER, near Boston. We received another from J. W. AUDUBON, who procured it at the falls of Niagara ; we have also frequently found it in the northern parts of New York, where the *Arvicola Pennsylvanica* likewise exists ; and we recently observed specimens near Detroit in Michigan. It appears, however, not to be found as far to the south as Wilson's Meadow-Mouse, as we have not succeeded in tracing it to the southern counties of Pennsylvania, where we have sought to obtain it.

GENERAL REMARKS.

We are not certain that this species may not have been indicated, although not accurately, by RAFFINESQUE in the American Monthly Magazine, under the name of *Lemmus noveboracensis*. His descriptions, however, in every department of natural history, are so short, vague, and imperfect, that it is impossible to identify his species with any degree of certainty ; they have created such confusion in the nomenclature that

Plate CXLIV

Fig 1.   Townsend's Arvicola.   Fig 2.   Sharp-nosed Arvicola   Fig 3.   Bank Rat.

Fig 1   Fig 2   Fig 3.

Drawn from Nature by J.W. Audubon.   On Stone by Wm E. Hitchcock.   Lith. Printed & Col.d by J.T. Bowen, Phil.

nearly all European and American naturalists have ceased to quote him as authority.

Sir JOHN RICHARDSON has described an Arvicola from the Rocky Mountains, which he refers to the above species (*A. noveboracensis*) of RAFFINESQUE, but which differs widely from the species here described.

ARVICOLA ORYZIVORA.—Bach.

RICE MEADOW-MOUSE.

PLATE CXLIV. Fig. 3

A. Pennsylvanicam equans, capite longo, rostro acuto, corpore gracili, auriculis prominulis, cauda longitudine trunci; supra ferruginea rufus, subtus subalbida.

CHARACTERS.

Size of Arvicola Pennsylvanica; *head, long; nose, sharp; body, slender; ears, prominent; tail, the length of the body, without including the head; colour, rusty brown above, beneath whitish.*

DESCRIPTION.

In form this species bears a distant resemblance to the cotton rat (*Sigmodon hispidum*); it is, however, a much smaller species. The ears, which are half the length of the head, are rounded, and are thickly clothed with hair on both surfaces; the feet are rather small; there is a short blunt nail in place of a thumb; under surface of palms, and tarsus, naked; toes on the hind-feet, long, the three middle ones of nearly an equal length; tail, rather long, thickly clothed on both surfaces with short hairs; whiskers, short, scarcely reaching the ears.

COLOUR.

The fur on the upper surface is slate colour, tipped with light brown and black, giving it on the dorsal aspect a dark grayish-brown tint, fading into lighter on the sides, and into whitish-gray on the belly and under surface; the ears are of the colour of the sides; feet, whitish; tail, brown on the upper surface, lighter beneath; whiskers, black and white.

DIMENSIONS.

	Inches.	Lines.
Length of head and body, - - - - - -	5	2
" of tail, - - - - - - -	4	

	Inches.	Lines.
From end of heel to point of longest nail, - - -	1	2
" point of nose to ear, - - - - -	1	
Height of ear, - - - - - - - -		5¾

HABITS.

The Rice Meadow-Mouse, as its name implies, is found in particular localities in the banks of the rice-fields of Carolina and Georgia. It burrows in the dykes or dams a few inches above the line of the usual rise of the water. Its burrow is seldom much beyond a foot in depth. It has a compact nest at the extremity, where it produces its young in April. They are usually four or five. In spring this Mouse is in the habit of sitting on the dams near the water, and is so immoveable, and so much resembles the colour of the surrounding earth, that it is seldom noticed until it moves off to its retreat in the banks. We have observed it scratching up the rice when newly planted and before it had been overflowed by the water. When the rice is in its milky state this animal commences feeding on it, and continues during the autumn and winter, gleaning the fields of the scattered grains. We have also seen its burrows in old banks on deserted rice-fields, and observed that it had been feeding on the large seeds of the Gama grass (*Tripsicum dactyloides*), and on those of the wild rye (*Elymus Virginicus*). A singular part of the history of the Rice Mouse is the fact that in the extensive salt-marshes along the borders of Ashley and Cooper rivers, this species is frequently found a quarter of a mile from the dry ground. Its nest is suspended on a bunch of interlaced marsh grass. In this situation we observed one with five young. At certain seasons this little animal feeds on the seeds of the marsh grass (*Spartina glabra*). When these fail it sometimes retires to the shore for food, but has no disrelish to the small crustacea and mollusks that remain on the mud at the subsiding of the tide.

This species swims rapidly, and dives in the manner of the European water-rat (*Arvicola amphibia*), or of our *Arvicola Pennsylvanica*. In an attempt at capturing some alive, they swam so actively, and dived so far from us, that the majority escaped. Those we kept in captivity produced young in May and September; they were fed on grains of various kinds, but always gave the preference to small pieces of meat.

GEOGRAPHICAL DISTRIBUTION.

We obtained several specimens of this Mouse through the aid of our friend Dr. ALEXANDER MOULTRIE, who assisted us in capturing them on his rice plantation in St. John's parish, South Carolina. We procured a considerable number on the salt marshes near Charleston, saw several on the eastern banks of the Savannah river, and near Savannah ; and the late Dr. LEITNER brought us a specimen obtained in the Everglades of Florida. This Arvicola is said to exist as far to the north as New Jersey.

GENERAL REMARKS.

We obtained specimens of *Arvicola Oryzivora* in the winter of 1816, but did not describe it until May 1836, when we designated it by the above name. Having occasion to send descriptions of several, then undescribed, species to the Academy of Natural Sciences of Philadelphia, we sent a specimen of this animal to Dr. PICKERING, requesting him and Dr. HARLAN to compare it with the *Arvicola riparia* of ORD, a species which we had not seen, stating our reasons why we regarded it as distinct. In searching in the Academy, a specimen of this species was found, and Dr. HARLAN, in opposition to the views of PICKERING, felt himself authorized to publish it in SILLIMAN's American Journal (vol. xxxi.), bestowing on it the name of *Mus palustris*, making use of the head of our specimen for an examination of the teeth.

The teeth and general appearance of this species, the form of its body, and especially its ears and tail being thickly clothed with hair, render it apparent that it does not belong to the genus Mus, but is more nearly allied to Arvicola. As the name " *Arvicola palustris*" is pre-occupied (HARLAN's Fauna, p. 136), we are favoured with an opportunity of extricating it from the confusion of synonymes in which it would otherwise be involved, and of restoring it to its true genus under the name given by its legitimate describer.

Drawn from Nature by J. W. Audubon.

On Stone by W. E. Hitchcock.

Lith Printed & Col'd by J. T. Bowen Phil.

Townsend's Shrew-Mole

SCALOPS TOWNSENDII.—Bach.

TOWNSEND'S SHREW-MOLE.

PLATE CXLV.—Males.

S. Magnitudine S. aquatico duplo major. supra rufo-fuscus. Dentibus XLIV.

CHARACTERS.

Double the size of the common Shrew-Mole, with eight more teeth than that species ; dark liver colour.

SYNONYMES.

COMMON MOLE. Mackenzie's Voyage to the Pacific, &c., p. 314.
MOLE. Lewis and Clark, Journey, vol. iii. p. 42.
SCALOPS CANADENSIS. Rich, Fauna Boreali Americana, p. 9.
"   TOWNSENDII. Bach., Jour. Acad. Nat. Sci., vol. viii., part 1, p. 58.

DESCRIPTION.

Dental Formula.—Incisive $\frac{2}{4}$; Canine $\frac{6-6}{6-6}$: Molar $\frac{4-4}{3-3}$ = 44.

In the upper jaw the incisors are large, and a third higher than the canine teeth usually termed false molars, which immediately follow them ; these are succeeded by three small teeth of a nearly conical shape, increasing in length from the first to the third ; the fourth false molar on each side is the smallest ; the fifth is a little larger in size, and slightly compressed ; the sixth still larger, and has a considerable posterior projection ; the four posterior cheek teeth, or true molars, are much larger and higher than the anterior ones ; the first of these (which we have called a canine tooth) is rather small, and bilobed, with a small internal tubercle ; the second and third are the largest and nearly resemble each other, exhibiting three distinct points, two external and posterior, and one anterior, the external ones being the longest, and the last molar being the smallest, and of a triangular form ; in the lower jaw there are two very small incisors in front ; next to these are two of a considerably larger size, which, although we have called them incisors, are nearly of the same shape and appearance as those which succeed them.

The canine or false molars, six on each side, are nearly the same size, and incline forwards ; the three true molars, which succeed, are large, nearly uniform in size, and correspond with those in the upper jaw, although they are smaller.

Body, thick and cylindrical, shaped like the Shrew-Mole (*Scalops aquaticus*) ; the limbs are short, being concealed by the skin of the body nearly down to the wrist and ankle-joints.

Palms, naked, very broad, furnished with moderately long nails which are channelled beneath ; tail, rather thick, tapering from the root to the tip, and nearly naked, being very sparingly clothed with short hairs ; the vertebræ are equally four sided ; fingers, very short, united to the roots of the nails ; nails, slightly curved ; hind-feet, more slender than the fore-feet, and distinctly webbed to the nails ; the feet are thinly clothed above, with short hairs. The whole of the body, both upper and lower surface, presents a velvety appearance.

COLOUR.

The body is dark liver brown colour above, changing with the light in which it is viewed to silvery or black shades ; the hair when blown aside exhibits a grayish-black colour to near the tips, which in some of the points are white, others brown black, producing the changeable colours above described. One of the specimens which we have seen—the one figured in our plate—has a whitish-yellow stripe about two lines wide, running in a somewhat irregular line along the under surface of the body to within an inch and a half of the insertion of the tail ; there is also a white streak commencing on the forehead, spreading over the snout and around the edges of the mouth and lower jaws. The teeth are white ; feet, point of nose, and tail, flesh colour ; nails, light brown.

DIMENSIONS.

							Inches.	Lines.
Length of head and body, -	-	-	-	-	-	8	6
" tail, -	-	-	-	-	-	-	1	6
Breadth of palm,	-	-	-	-	-	-		7

HABITS.

We were informed by Nuttall and Townsend, who mistook this species for our common Shrew-Mole (*Scalops aquaticus*), that they dug and formed galleries, and threw up little mounds of earth precisely in the manner of that animal. They are well known to the farmers and settlers in the valleys of Oregon, as they traverse their fields and gardens, cutting up the ground in some places to an injurious extent.

GEOGRAPHICAL DISTRIBUTION.

This species is found in considerable abundance near the banks of the Columbia and other rivers in Oregon, where our specimens were obtained. We are unable to say what is the northern limit of this animal. It has not yet been found on the eastern side of the Rocky Mountains, and we have not been able to determine positively that it exists in California; but we have little doubt that it is the most common Shrew-Mole on the Pacific side of the North American continent, where our common species (*Scalops aquaticus*) does not appear to have been discovered.

GENERAL REMARKS.

Sir JOHN RICHARDSON, who first described this animal from a specimen preserved in the museum of the Hon. Hudson's Bay Company, obtained by Mr. DAVID DOUGLAS, does not seem to have made a comparison between this Mole and our common Atlantic species. HARLAN had described the skull of the species which we have since described and figured as *Scalops Brewerii*, having forty-four teeth, and another which had thirty-six. RICHARDSON was thus induced to suppose that authors had varied in their descriptions of the *Scalops* from their having mentioned edentate spaces between the incisors and grinders, and had consequently described the young in those specimens which had only thirty-six teeth. The young, however, of our common *aquaticus* (or as CUVIER has called it, *Scalops Canadensis*) has only thirty teeth, the adult thirty-six, whilst the present species has forty-four.

On our pointing out to Sir JOHN RICHARDSON these particulars, he expressed himself gratified to have an opportunity of correcting the error into which he had inadvertently fallen.

GENUS DASYPUS.—Linn.

DENTAL FORMULA.

Incisive $\frac{0}{0}$ *or* $\frac{2}{4}$; *Canine* $\frac{0-0}{0-0}$; *Molars varying in the several species from* 28 *to* 68 ; *these teeth cylindrical, separate, and without enamel on the inner side.*

Head, long ; mouth, small ; tongue, partially extensible. Body, altogether covered with a shell, or plate armour. Four or five toes to the forefeet, five toes to the hind-feet. Toes, armed with long nails for digging ; mammæ, two or four. Tail, rather long, round.

Stomach, simple ; intestines, without cæca.

Habit, living in woods, on ants, roots, and putrid animals ; rolling themselves up for protection ; confined to the warmer parts of America.

Nine species belonging to this genus have been described by authors.

The genus requires a revision, and the species will no doubt, from the rage which exists at present for making new genera, be greatly subdivided.

The generic appellation is derived from δασυς, *dasus*, rough, and πους, *pous*, a foot.

DASYPUS PEBA.—Desm.

NINE-BANDED ARMADILLO.

PLATE CXLVI.—Male.

D. Dentibus primoribus laniariisque nullis, molaribus $\frac{8-8}{8-8} = 32$, cauda tereti, cingulis circumdata, ad apicem solum nuda, testa zonis mobilibus, auriculis longissimis.

CHARACTERS.

No incisive or canine teeth ; Molars $\frac{8-8}{8-8} = 32.$

Tail, round, with rings nearly its whole length. Body, with mobile bands ; ears, very long.

SYNONYMES.

Dasypus Peba. Desm., Mammal., p. 368.
    "   Septem Cinctus, D. Octo Cinctus, and D. Novem Cinctus. Linn.

ARMADILLO BRAZILIANUS. Briss., Regne Animal, 40.
     "   MEXICANUS.   "   "   **41.**
     "   GUYANENSIS.   "   "   **42.**

CACHICAME. Buffon, Hist. Nat., x. p. 250.

TATOU NOIR. D'Azara, Paraguay, vol. ii. p. 175.

TATU PEPA. Marc., Brazil, 231.

NINE, EIGHT, OR SEVEN BANDED ARMADILLO. Pennant's Quadrupeds, Synopsis, pp. 324, 253.

PIG-HEADED ARMADILLO. Grew, Mus, p. 19, t. i.

SIX-BANDED ARMADILLO. Shaw's General Zoology, vol. i., part 1, p. 189.

GÜRTELTHIER MIT ACHTZEHN GÜRTELN. Schreb., pp. 227, 228.

DESCRIPTION.

This singular production of nature, it might be said, resembles a small pig saddled with the shell of a turtle ; it is about the size of a large opossum ; the head is small, and greatly elongated, and the neck can be retracted so far as to entirely withdraw the head under the shell. Muzzle, narrow and pointed ; mouth, large : tongue, aculeated, and can be drawn out three inches beyond the nose.

The head and nose are covered with rather small plates irregularly shaped, most of them hexagonal. There are on the back nine transverse bands in the specimen from which we describe, although the number of bands is occasionally only seven or eight. The shoulders, hams, and rump are protected by two plates, covered with large scales regularly arranged in distinct rows following the direction of the movable transverse bands, and descending lower towards the ground than the bands, forming a sort of flap over the shoulders and over the hips like the skirt of a saddle. Thus the covering of the head may be compared to a helmet, and that of the shoulders and on the hind parts to breast-plates and thigh-pieces, the whole forming an almost impenetrable coat-of-mail.

The tail is protected by numerous rings, furnished with scales of the same substance, shape, and hardness, as those on other parts of the upper surface of the body. The texture of this shell-like covering of the Armadillo appears to be something between turtle-shell or horn, and very hard sole-leather. The eyes are small, and placed far back in the head, on a line with the corner of the mouth.

Legs, short and stout ; nails, strong, sharp, very slightly hooked, and not channelled beneath ; there are four toes on each fore-foot, the middle ones being much the longest, and the outer, shorter, and situated far behind ; there are five toes on the hind-feet, the central being longest, the first and fifth shortest, and the two others nearly of an equal length. Ears, long, narrow, and pointed, destitute of hair, and the skin on their upper surface slightly granulated, but not protected by scales. The under surface of the body is only covered by a soft leathery skin, as also the legs ; the front of each foot is protected by scales for about two inches above the toes.

A few scattered hairs can be observed on the under surface of the body, and here and there a single hair along the edges of the plates above ; the animal may nevertheless be described as hairless. Mammæ, four.

COLOUR.

Entire surface of body, ochreous brownish-yellow ; browner along the sides of the head and beneath the ears ; feet and nails, yellowish-brown.

DIMENSIONS.

	Foot.	Inches.
From point of nose to root of tail, - - - -	1	6
Tail, - - - - - - - - -		8
Height of ear, - - - - - - - -		2
Point of nose to eye, - - - - - -		$2\frac{7}{8}$
Nose to ear, - - - - - - -		$4\frac{1}{2}$
Longest nail on fore-foot, - - - - - -		1
"　　" on hind-foot, - - - - -		$\frac{7}{8}$

HABITS.

The Armadillo is not "a fighting character," but on the contrary is more peaceable than even the opossum, which will at times bite in a sly and treacherous manner, quite severely. Indeed nature, whilst giving to the Armadillo a covering of horn-plates or scales, which serve to protect it from many of its foes, has not supplied it, as she has other non-combating animals—the porcupine for instance—with sharp-pointed quills or spines, and its only means of aggression are its claws, which although large are better adapted for digging than aught else. The animal, how-ever, sometimes has been known when caught by the tail, to kick rather hard with both fore and hind-legs, so that its captor was glad to let go, for it possesses great strength in the limbs. A friend of ours who formerly resided in South America had a pet Armadillo in his bed-chamber, where it generally remained quiet during the day, but in the dark hours was active and playful. One night after he had gone to bed, the Armadillo began dragging about the chairs and some boxes that were placed around the room, and continued so busily engaged at this occupation that our friend could not sleep. He at length arose and struck a light, when to his surprise he found boxes he had supposed greatly too heavy for such an animal to stir, had been moved and placed together so as to form a sort of den or hiding-place in a corner, into which the animal retreated with great apparent satisfaction, and from whence it could only be drawn out after a hard struggle, and the receipt of some severe strokes from its claws. But in general the Armadillo does not evince any disposition to resent an attack, and in fact one of them when teased by a pet parrot, struck out with its claws only till pressed by the bird, when it drew in its head and feet, and secure in its tough shell, yielded without seeming to care much about it, to its noisy and mischievous tormentor, until the parrot left it to seek some less apathetic and more vulnerable object to worry.

But when the Armadillo has a chance of escape by digging into the ground, it is no sluggard in its movements, and progresses towards the depths of the soil with surprising rapidity. This animal however on being much alarmed rolls itself up, and does not attempt to fly, and it is chiefly when it has been digging, and is at or near the mouth of a hole, that it tries to escape ; preferring generally, to be kicked, tumbled about with a stick, or be bitten at by a dog, to making an effort to run.

We have heard it asserted that when it has the advantage of being on a hill or elevated spot, the Armadillo upon the approach of danger, forms a ball-shaped mass of its body, with the tail doubled under the belly, starts down the hill and rolls to the bottom.

The principal food of this genus consists of ants of various species, which are so abundant in some portions of Central and South America as to be great pests to the inhabitants of those parts of the world. A large species of this family, however (*Dasypus giganteus*), is described by D'Azara as feeding on the carcases of dead animals ; and it appears that in neighbourhoods where that Armadillo is found, the graves of the dead are protected by strong double boards, to prevent the animal from penetrating, and devouring the bodies. Armadillos are said to eat young birds, eggs, snakes, lizards, &c. It should perhaps here be remarked that the large Armadillo just mentioned, although covered with plates or scales like our present species (*D. peba*), and similar in form, is very different in its organization, and has indeed been characterized by F. Cuvier under the new genus *Priodontis*.

To return to our present species. The Nine-banded Armadillo is, as we were informed by Captain Charles H. Baldwin, kept in Nicaragua, not only by the people of the ranchos, but by the inhabitants of some of the little towns, to free their houses from ants, which, as is said, it can follow by the smell. When searching for ants about a house, the animal puts out the tongue and scrapes the ants into the mouth from around the posts on which the houses are raised a little above the ground, and has been known to dig down under the floors, and remain absent for three or four weeks at a time.

When burrowing this species utters a slight squeak, quite faint however. They are said to dig down in a straight direction when they discover a subterranean colony of ants, without beginning at the mouth or entrance to the ant-hole. There are two favourite species of ant with the Armadillo in Nicaragua, one of which makes nests in the forks of trees in the forests. The tree ants are white, the others small and black. The Armadillos keep about the roots of the trees in order to feed upon the former, and as we have already said, dig for the latter. They also root up the ground with their pig-like snout, and do some damage to gardens. They are very persevering when in pursuit of ants, and whilst they turn up the light soil with the snout, keep the tongue busy taking in the insects.

It has been assured us that when a line of ants (which may sometimes extend some distance in the woods) are busily engaged in carrying provision to the general storehouse, they scatter in every direction at the instant the Armadillo begins to dig down towards their stronghold, evidently having some communication from head-quarters equivalent to "sauve qui peut."

The gait of these animals when not alarmed is like that of a tortoise, and about as fast. They have nails powerfully organized for digging, whilst their legs are only long enough to raise the body from the ground. The holes the Armadillo excavates in the earth for its own purposes, are generally dug at an angle of forty-five degrees, are winding, and from six to eight feet long.

The Armadillo is generally much darker in colour than the specimen we figured, which having been a pet, was washed and clean when we drew it. When in the woods these animals partake more or less of the colour of the soil in which they find their food, as some of the dirt sticks to their shell. Those that have been domesticated prefer sleeping above ground, but this animal when wild lives in burrows, holes in the roots of trees, or under rocks.

From our esteemed friend Capt. J. P. McCown, U. S. A., we have the following : "The Armadillo is to be found in the chaparals on the Rio Grande. I have seen their shells or coat-of mail on the prairies ; whether carried there by larger animals, or birds, or whether they inhabit the prairies, I cannot say. I have seen many that were kept as pets and appeared quite tame. I am inclined to the opinion that there are two species—the larger living on the low and wet lands and in the canebrakes, the smaller occupying the rocky hills and cliffs."

Plate CXLVI

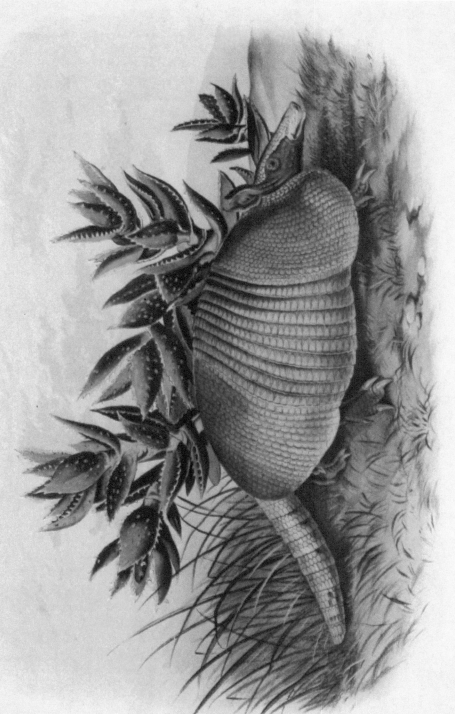

Drawn from Nature by J.W. Audubon

On Stone by Wm E. Hitchcock

Lith. Printed & Col.d by J.T. Bowen, Phil.

Nine banded Armadillo

This animal is said to produce three or four young at a time. Its flesh is eaten by the Spaniards and natives. It has been described to us by Americans who ate of it during the Mexican war, to be about equal to the meat of the opossum ; we have heard, however, from South Americans, that it is considered quite a delicacy, being white, juicy, and tender ; it is cooked by roasting it in the shell.

The South American negroes catch the Armadillo at night. When they are in the woods their dogs scent the animal and run it to its hole (if it be near enough to its retreat to reach it). It is then dug out by the blacks, although sometimes known to excavate its burrow to a considerable depth below its usual place of rest, whilst the diggers are at work after it. Two or three of these animals generally keep together, or near each other, and the negroes always expect to kill more, when they have captured one. They are said to run pretty fast when trying to reach their holes, but the manner of their gait at such times is not known to us. Their holes are often dug in the sides of steep banks or hills, and in thick and dense parts of the woods.

We have heard that in some parts of Nicaragua the Armadillos are so common that they can be purchased for a *medio*—six and a quarter cent piece.

GEOGRAPHICAL DISTRIBUTION.

This animal is described as existing in Brazil in South America : it is found in Guiana and Central America, is common in Mexico, and is found in the southern portions of Texas. It is not very uncommon near the lower shores of the Rio Grande.

GENERAL REMARKS.

It is stated that another species of Armadillo inhabits the northern part of Mexico and penetrates also into Texas. Thus far, however, we have been unable to detect any other species than the present as having been seen within the geographical limits to which this work has been restricted.

It is now ascertained that the number of bands on the Armadillo forms no safe guide in designating the species, inasmuch as the bands vary in different individuals of the same species, and D'AZARA, moreover, has shown that there are individuals of different species which have the same number of bands.

SPERMOPHILUS TOWNSENDII.—Bach.

AMERICAN SOUSLIK.

PLATE CXLVII. Fig. 1.—Male.

S. Magnitudine Sciuri Hudsonii, capite parvo, corpore gracilior, auribus caudaque brevibus, colore supra rufo-fusco griseo sparsim vario, infra pallidiore.

CHARACTERS.

Size of Sciurus Hudsonius (*red squirrel*); *head, small; body, rather slender; ears and tail, short; colour, upper surface speckled with white and brown; beneath, yellowish-gray.*

SYNONYMES.

Arctomys (Spermophilus) Guttatus ?—American Souslik. Rich., F. B. A., p. 162.
Spermophilus Townsendii—Townsend's Marmot. Aud. and Bach., Jour. Acad. Nat.
Sci. Phil., vol. viii., part 1, p. 61.

DESCRIPTION.

This animal has a convex and obtuse nose, with the frontal bone depressed; the body is rather long and slender; head, short; ears, slightly visible above the fur; cheek pouches, small; nails, slender, compressed, and slightly arched; the thumb protected by an acute and prominent nail; the second toe of the fore-foot, as in all the species of the genus, is longest, and not the third, as in the squirrels. The first toe is a little shorter than the second, and the third intermediate in length between the first and second. The tail appears (in the dried specimen) much flattened; it is clothed with hairs which are longest on the sides.

The fur is throughout remarkably soft, smooth, and lustrous.

COLOUR.

There is a line of white around the eye. The fur on the whole upper surface is, for one fourth of its length from the roots, dark bluish, or nearly black, then (a broad line of) silver gray, then (a narrow line of) dark brown edged with yellowish-white, giving it a brownish-gray appearance, speckled with white all over the back; these spots are longest near the dorsal line, becoming smaller half way down the sides. An indistinct line of separation between the colours of the upper and under surfaces appears high up along the hips and sides; on the under surface, the hairs are nearly black at the roots, and are cinereous at the tips; on the forehead, nose, and sides of the neck, there is a slight tinge of light yellowish-brown. Tail, on the upper surface, light yellowish-brown edged with whitish; beneath, whitish, with a slight tinge of brown; teeth, white; nails, black.

DIMENSIONS.

							Inches.	Lines.
From point of nose to root of tail,	-	-	-	-	8	9
Head, -	-	-	-	-	-	-	1	10
Tail (vertebræ), -	-	-	-	-	-	1	
" (to end of fur),	-	-	-	-	-	1	6
Length of heel to end of middle claw,	-	-	-	1	4

HABITS.

In a letter addressed to us by the late Mr. TOWNSEND he states that this handsome Spermophile, in summer inhabits the prairies near the Wallawalla, where it is rather common: it becomes excessively fat, and is eaten by the Indians. It disappears in August and re-appears early in spring in a very emaciated state. We have heard from other sources that it lives in small families, like the Spermophiles, generally burrowing in holes, and that it is seen either sitting on the side of them or with the head partially protruded, but disappears in its underground retreat, on the approach of man or any other animal.

GEOGRAPHICAL DISTRIBUTION.

This species exists on the western sides of the Rocky Mountains in Oregon, where the few specimens we have seen have been obtained.

GENERAL REMARKS.

RICHARDSON described this species under the name of *A. guttatus*, an animal described by PALLAS (Glir. tab. 6 B) existing on the Wolga in Russia; but BUFFON mentions of that species, that the name of Souslik is intended to express the great avidity that animal has for salt, which induces it to go on board vessels laden with that commodity, when it is often taken. We should judge that its American relative has less opportunity of indulging in such a propensity. We carried a specimen with us to Europe, and had an opportunity at the Berlin Museum of comparing it with specimens from Siberia ; there is a general resemblance between the animals of the two countries, but they are scarcely more alike than the red squirrel of Europe (*Sciurus vulgaris*) and the red squirrel of America (*Sciurus Hudsonius*). They may be distinguished from each other at a glance by the large rounded spots on the back of the Russian animal, compared with the white and irregular specks in the American species.

As the name *guttatus* was pre-occupied, we have named this animal anew, and in doing so, called it after the gentleman who furnished us the specimen.

ARVICOLA TEXIANA.—Aud. and Bach.

TEXAN MEADOW-MOUSE.

PLATE CXLVII. Fig. 2.—Male.

A Sigmodon hispidum minore, supra rufo-fuscus nigro sparsim notato, striis nigris lateralis, lateribus fuscus, infra albido.

CHARACTERS.

Smaller than the cotton rat (Sigmodon hispidum) ; *back, brownish-yellow, spotted with irregular small blotches of black, a faint obscure stripe of black on each side. Sides, reddish-brown ; belly. whitish-gray.*

DESCRIPTION.

This new species bears a general resemblance to the cotton rat of South Carolina and Georgia. Head, of moderate size ; body, rather slender, and thin in the flanks ; hair, soft ; under fur, woolly ; ears, large, ovate in shape, extending beyond the fur, and nearly naked behind, with, on the margins, a few scattered hairs. Whiskers, numerous, and about as long as the head ; four toes on the fore-feet, with a small and almost imperceptible nail in place of a thumb. Five toes on the hind-feet, the outer and inner of nearly equal length, the other three longer, and each of about the same length ; legs, slender ; feet, covered by short hairs, and with a few hairs between the toes, not however concealing the nails ; heel, narrow and naked, as is also the under surface of the fore-feet.

Tail, rather long and slender, tapering to the point, and thinly covered with hairs.

COLOUR.

Fur, dark slate colour on the back at the roots, with reddish-yellow tips. The longer hairs, which are fine and soft, are irregularly marked with dark and yellowish-white at the base, and tipped with dark brown to where they mingle with the under fur. When the hair is laid smooth there is an obscure black stripe on the sides of the back, running from behind the shoulders towards the rump and converging across the buttocks to a point

at the insertion of the tail ; the remainder of the back between these
stripes is somewhat irregularly, and very slightly, waved or barred as it
were, with dark brown spots on a yellowish ground ; head, yellowish-
brown ; sides of the neck, and along the flanks to the hip, brownish-yellow.
A narrow line of yellowish-white extends under the chin and on the belly ;
tail, brown above, grayish-white beneath ; ears, brownish yellow ; whiskers,
white, with a few brown hairs interspersed.

DIMENSIONS.

	Inches	Lines.
From point of nose to root of tail,	4	7
" " to ear,	1	3
" " to eye,		6
Length of tail,	4	
From heel to point of longest claw,	1	3

HABITS.

This is an active and rather pugnacious little rat. It is sometimes to
be seen near the edges of the chaparals, in which it makes its nest. It
mostly feeds on seeds of wild grains and grasses, although it has recently
shown a disposition to frequent the farm-yards of our enterprising Texan
settlers.   Like the Arvicolæ generally, this animal is a good swimmer, and
takes the water when the rains flood the flat plains, which it has pleased
the Texans to denominate " hog-wallow prairies."   In the spring season
this rat devours a good many eggs of such small birds as make their nests
on the ground or in the rank grass and weeds, and it does not hesitate to
eat any dead bird or small animal it may find.

Not being very numerous, it is difficult to procure it, and as setting traps
for small animals, baited with meat, in the chaparal, has been found almost
useless owing to larger quadrupeds than those intended to be caught seizing
the flesh, and breaking the trap to pieces, or (as is often the case) devouring
the small ones that may have been already entrapped, there is no probability that this or other small species which inhabit Texas and the neighbouring countries will become familiar objects in our collections of mammalia for some time to come.   We have therefore placed our specimens of
this rat in the museum of the Charleston College at Charleston, South
Carolina, where may also be found the skins of some other animals first
described in our work, and of which specimens, so far as we have heard,
have not been procured by others.

GEOGRAPHICAL DISTRIBUTION.

This species was first discovered on the river Brasos, and afterwards seen in the country along the Nueces and Rio Grande, where chapparal thickets afford it shelter

GENERAL REMARKS.

Although this Meadow-Mouse approaches nearer to our cotton rat than any other Arvicola with which we are acquainted, it presents very striking differences ; its form is lighter and more slender, its heel narrower, tail proportionably longer, and fur much softer. The cotton rat is of a uniform colour on the back, except the ends of the long hairs being tipped with white, but the Texan species presents a somewhat indistinct appearance of specks or spots of blackish on a yellow ground.

ARVICOLA OREGONI.—Bach.

Oregon Meadow-Mouse.

PLATE CXLVII. Fig. 3.—Male.

A. M. musculus magnitudine, gracilior, auribus brevibus vellere abscon‐
ditis ; colore supra cinereo fusco, subtus cinereo.

CHARACTERS.

*Size of the house mouse ; slender form ; ears, short, nearly naked, and con‐
cealed by the fur ; colour, ashy brown on the back, cinereous beneath.*

SYNONYME.

Arvicola Oregoni—Oregon Meadow-Mouse. Bachman, Jour. Acad. Nat. Sci.,
vol. viii., part i., p. 60.

DESCRIPTION.

Head, of moderate size ; body, slender, eyes, small ; ears, concealed by
the fur and nearly naked, being clothed with but a few short and scattered
hairs ; feet, small ; whiskers, as long as the head ; a very minute blunt
nail on the fore-foot ; tail, and feet, clothed with short hairs.

COLOUR.

The fur, on the upper surface, is bluish from the roots to near the tips,
where most of the points of the hairs are black, a shade of brownish
appearing beneath ; under surface, ashy white ; above the eye, and imme‐
diately in front of the shoulders there is a line of light brown ; whiskers,
white and black, the latter predominating ; feet, flesh coloured ; incisors,
yellow ; tail, dark brown above, yellowish-white beneath, with the extreme
end black. In Townsend's Notes it is stated that the specimen we have
described above, was an old male, captured on the 2d of November, 1836.

DIMENSIONS.

	Inches.	Lines.
Length of head and body, - - - - - -	3	0
" of tail, - - - - - - - -	1	2

Plate CXLVII

Fig.1

Fig.2

Fig.3

On Stone by Wᵐ E. Hitchcock

Fig.1 American Sorelek   Fig.2 Oregon Meadow Mouse   Fig.3 Texan Meadow Mouse

Drawn from Nature by J.J. Audubon.

Lith. Printed & Col.ᵈ by J.T. Bowen, Phil.ᵃ

HABITS.

We are unacquainted with the habits of this species, but should judge from its form resembling that of LE CONTE's pine mouse (*A. pinetorum*) that instead of having galleries on the surface of the earth. as is the general habit of the Meadow-Mice, this species lives principally under ground. and only comes to the surface at night. to seek its food ; it evidently feeds more on roots than on seeds.

GEOGRAPHICAL DISTRIBUTION.

This Arvicola was captured in Oregon, near the Columbia river.

GENERAL REMARKS.

Although its head is rather smaller in proportion than that of LE CONTE's pine mouse (*Arvicola pinetorum*), and its body differing from that animal in colour. this species is nevertheless very similar to it in form. more especially in the almost naked lobe of the ear ; and seems to be the representative of that Atlantic species in Oregon.

PUTORIUS FUSCUS.—Aud. and Bach.

Tawny Weasel.

PLATE CXLVIII. Male.

P. Corpore inter putorius erminius et P. vulgaris intermedio ; caudâ illius breviore, sed hujus longiore ; apice nigro ; vellere supra fusco ; subtus albo.

CHARACTERS.

Intermediate in size between the ermine and the common weasel of Europe ; tail, shorter than in the former, but longer than in the latter, with the extremity black ; body, brown above, white beneath.

SYNONYMES.

Mustela Fusca. Aud. and Bach., Jour. Acad. Nat. Sci. Phil., October 5, 1841, p. 94.
    "   "   DeKay, Nat. Hist. State of New York, p. 35.

DESCRIPTION.

Body and neck, rather short in proportion to others of this genus, and far more robust than the common European Weasel. The feet especially appear a third larger, and are more thickly clothed with fur, which covers the palms and toes, and conceals the nails completely ; ears, a little longer and more pointed than those of either the ermine or common Weasel.

In writing this description we have several specimens of the European common Weasel (*P. vulgaris*) before us, and the ends of the tails in that species are uniformly brown, with here and there a black hair interspersed. Although the hair of the present species is black at the extremity of the tail, like that of the ermine, yet these hairs are short and soft, and more like long fur, and do not present the long and coarse appearance of those of the latter species, but lie closer along the vertebræ, and form a sharp point at the extremity.

Claws, short and stout ; incisors equally large with those of the ermine, but shorter ; ears, large, obtusely pointed at tip, and thinly clothed with short adpressed hairs ; tail, cylindrical, and narrowed down to a point of fine hairs, the tip somewhat resembling a large water-colour pencil or brush. Whiskers, as long as the head, and rather numerous. The hairs on the

Plate CXLVIII

Drawn from Nature by J. W. Audubon

On Stone by Wm E. Hitchcock

Lith. Printed & Col.d by J. T. Bowen Phil.

Tawny Weasel

body are of two kinds : the longer hairs are a little more rigid, and far more numerous, than on the ermine, and the under fur is a little longer, coarser, and less woolly than the fur of the latter animal.

COLOUR.

The whole upper surface, sides, outside of legs, feet, ears, and tail to within an inch of the extremity, uniform tawny brown, except on the centre of the back and top of the tail, where the colouring darkens. Thus the body of the animal is a shade darker than the summer colour of the ermine, while the colour of the tail is, for about an inch, nearly as black as in that species. The white on the lower surface is not mixed with brown hairs as in *Putorius vulgaris*, and not only occupies a broader space on the belly, but extends along the inner surface of the thighs as low as the tarsus, whilst in *P. vulgaris* the white scarcely reaches the thighs. The whole of the under surface is pure white ; this colour does not commence on the upper lip, as is generally the case in the ermine, but on the chin, extending around the edges of the mouth, and by a well defined line along the neck, inner parts of the fore-legs, and inner parts of the thighs, tapering off to a point nearly opposite the heel on the hind-legs.

Whiskers, dark-brown, with a few white ones interspersed. The specimen from which our figure was drawn, was captured on Long Island in May 1834, and is therefore in summer pelage.

DIMENSIONS.

For the sake of convenient comparison we will also here give the dimensions of the two species of Weasel to which our animal is most nearly allied, taken from specimens now before us.

	P. fuscus.		*P. erminea.*		*P. vulgaris.*	
	Inches.	Lines.	Inches.	Lines.	Inches.	Lines.
Length of head and body, -	9	0	11	7	7	0
" tail (vertebræ), -	2	9	4	6	1	9
" " (including fur),	3	2	6	2	2	1
Height of ear posteriorly, -		3		2½		2

HABITS.

We find from our notes, that in the State of New York in the winter of 1808, we kept a Weasel, which we suppose may have been this species, in confinement, together with several young ermines. The latter all became white in winter, but the former underwent no change in colour, remaining brown. On another occasion a specimen of a brown Weasel was brought to us in the month of December. At that season the ermines are invariably white. We cannot after the lapse of so many years say with certainty whether these specimens of Weasels that were brown in winter were those of the smaller, *Putorius pusillus*, or the present species ; although we believe from our recollection of the size they were the latter. We therefore feel almost warranted in saying that this species does not change colour in winter.

We were in the habit of substituting our American Weasels for the European ferrets, in driving out the gray rabbit (*Lepus sylvaticus*) from the holes to which that species usually resorts in the northern States, when pursued by dogs (see vol. i. p. 59). Whilst the ermines seemed to relish this amusement vastly, the brown Weasel refused to enter the holes, and we concluded that the latter was the least courageous animal.

On one occasion we saw six or seven young Weasels dug out by dogs from under the roots of a tree in a swamp, which we believe to have been of this species.

GEOGRAPHICAL DISTRIBUTION.

The specimens which we have seen of this animal all came from different parts of the State of New York. We have however heard of the existence of a Weasel which is brown in winter in the States of Ohio and Michigan, which we have reason to believe is the present species.

GENERAL REMARKS.

Our early writers on natural history were under the impression that we had but one, or at farthest two species of Weasel in our country.

GODMAN supposed that there was but one Weasel in North America, and that it was the common Weasel in summer, but was the ermine in summer pelage, turning white in winter. HARLAN gave DESMAREST's description of the European *Mustela vulgaris*, supposing that animal to exist in our country.

RICHARDSON gave two species as belonging to North America, one of which he supposed to be identical with the common Weasel of Europe. It is now ascertained that we have at least five species in the United States, four of which are found in the State of New York.

SCIURUS FRÉMONTI.—Townsend.

FRÉMONT'S SQUIRREL.

PLATE CXLIX.—Fig. 1.

Magnitudine Sciuri Hudsonii ; caudâ corpore breviore ; auribus cris
tatis ; colore supra albido, infra cinereo.

CHARACTERS.

Size of Sciurus Hudsonius ; *tail, shorter than the body ; ears, tufted.
Colour, light gray above, ashy white beneath.*

DESCRIPTION.

Upper incisors, larger than those of *S. Richardsonii* or *S. lanuginosus ;*
lower incisors, longer and more curved than those of *S. Hudsonius.* The
first or deciduous tooth wanting.

Body, short and stout, presenting less appearance of lightness and agility
than that of the Hudson's Bay Squirrel ; head, short and broad ; forehead,
but slightly arched ; ears, rather short, broad, rounded, and much tufted ;
whiskers, long, reaching to the shoulders ; legs, short and stout ; the third
toe on the fore-foot, slightly the longest ; nails, compressed, and shorter,
blunter, and less hooked than those of *S. Hudsonius.* Tail, a little shorter
than the body, of tolerable breadth, and capable of a distichous arrangement.

The whole body is clothed with a dense coat of rather long and soft fur.

COLOUR.

Fur on the back, dark plumbeous from the roots ; on the sides, tipped
with light gray. There is a narrow dark reddish line along the centre of
the back, caused by the hairs on the dorsal line being tipped with reddish-
brown and black. On the under surface the fur is plumbeous at the roots,
and tipped with ashy white. The tufts on the ears are black ; whiskers,
black ; a line of dark brown runs from the end of the nose, blending gradually with the lighter tint of the forehead ; there is a light circle around
the eye ; sides of the nose, and lips, yellowish-white ; upper surface of feet,
gray.

There is a slight and almost imperceptible black stripe about a line wide and three inches long, separating the colour of the sides from the ashy white tint of the under surface. The annulations in the hairs of the tail are somewhat indistinct: from the roots for nearly half their length they are grayish-white, are then black, and are broadly tipped with white.

DIMENSIONS.

	Inches.
Length of head and body,	7
" tail (vertebræ),	4¾
" " (including fur),	6¼
Height of ear posteriorly,	⅜
" " (including tufts),	⅞
Palm and middle fore-claw,	1⅛
Sole and middle hind-claw,	2

HABITS.

We possess no information in regard to this animal farther than that it was obtained on the Rocky Mountains.

It no doubt, like all the other small species of Squirrels which are closely allied with it (*Richardsonii, Hudsonius, lanuginosus,* &c.), feeds on the seeds of pines, and other coniferæ.

All these squirrels inhabit elevated regions of country, and in addition to their habit of climbing, have burrows in the ground, wherein they make their dormitories, and dwell in winter; whilst in summer they select the hollow of a tree, in which they construct their nests.

Their note is peculiar, like *chicharee chicharee* repeated in quick succession, and differing from the *qua qua quah* note of the larger squirrels.

By their habit of burrowing or living in holes in the ground, these small squirrels make an approach to the genus *Tamias,* or ground squirrels.

GEOGRAPHICAL DISTRIBUTION.

The only specimen we have seen was obtained by Colonel FRÉMONT; it was procured on the Rocky Mountains, on his route by the south pass to California.

GENERAL REMARKS.

The tufts on the ears of this species are considerably larger than in any other known species of squirrel in our country, except *Sciurus dorsalis,* a beautiful new squirrel discovered in California by Mr. WOODHOUSE, and recently described by that gentleman, and in this respect bear a resemblance to those on the ears of the common Squirrel of Europe (*Sciurus vulgaris*); the tufts, however, of the latter are twice the length of those of *S. Frémonti*, being an inch long, whilst in the latter they are half an inch in length.

These tufts, in the specimen, originate on the outer surface of the ear, near the base, and the edges of the ear are only covered with short hairs, whilst in the European species not only the posterior portions, but also the upper edges, or rims of the ear, are thickly haired. producing so large and thick a tuft that the animal at first sight appears to have an ear more than an inch long.

SCIURUS FULIGINOSUS.—Bach.

SOOTY SQUIRREL.

PLATE CXLIX. Fig. 2.

Sciuro Hudsonio paullo major ; caudâ nonnihil planâ, et corpore multo breviore ; colore plerumque supra nigro, subfusco-flavo variegato ; infra subfusco.

CHARACTERS.

A little larger than the Hudson's Bay Squirrel (S. Hudsonius) ; *tail, flattish, and much shorter than the body ; general colour, black above, grizzled with brownish-yellow ; beneath, brownish.*

SYNONYME.

SCIURUS FULIGINOSUS. Bach., Monograph of the genus Sciurus, Trans. Zool. Soc., London, August, 1838.

DESCRIPTION.

Head, short, and broad ; nose, very obtuse ; ears, short, and rounded, slightly clothed with hair ; feet and claws, rather short and strong ; tail, short, and flattened, but not broad, resembling that of *Sciurus Hudsonius ;* the form of the body is like that of the Carolina gray Squirrel.

COLOUR.

The limbs externally, and feet, are black, obscurely grizzled with brownish-yellow ; on the under parts, with the exception of the chin and throat, which are grayish, the hairs are annulated with brownish-orange and black ; at the roots, they are grayish-white ; the prevailing colour of the tail is black above, the hairs being brown at the base, some of them obscurely annulated with brown, and at the apex pale brown ; on the under side of the tail, the hairs exhibit pale yellowish-brown annulations.

DIMENSIONS.

						Inches.	Lines
Length of head and body,	-	-	-	-	-	10	
" tail (vertebræ),	-	-	-	-	-	6	9

On Stone by Wᵐ E Hitchcock

Fig 1 Fremonts Squirrel. Fig 2 Sooty Squirrel.

Drawn from Nature by J W Audubon

Lith Printed & Colᵈ by J T Bowen Phil

		Inches.	Lines.
Length of tail (including fur), - - - - -		8	6
" palm to point of middle fore-claw, - -		1	8
" heel, to point of longest nail, - -		2	1
Height of ear posteriorly, - - - - - -			4
Length of fur on the back, - - - - -			7

HABITS.

This dusky looking species is found in low swampy situations, and is said to be very abundant in favourable localities.

During high freshets, when the swamps are overflowed to the height of several feet, they are very active among the trees, leaping from branch to branch, indifferent about the waters beneath. They feed chiefly on pecan nuts, and are deemed by the French inhabitants of Louisiana to be the most savoury of all the Squirrels.

GEOGRAPHICAL DISTRIBUTION.

We have heard of this species as existing only in Louisiana and Mississippi, and as being chiefly confined to the swamps.

GENERAL REMARKS.

We are under the impression that this Squirrel is subject to considerable variations in colour. We obtained, through the kindness of Col. WADE HAMPTON, a number of specimens of the different Squirrels existing along the shores of the Mississippi, and among them we found several examples of this species. Some of them were of much lighter colours than the one which we described. In Louisiana, they are often so dark in colour, as to be called by the French inhabitants *le petit noir*.

The specimen from which our original description was made, was procured near New Orleans, on the 24th of March, 1837. It agrees in many particulars with a skin deposited in the late museum of Mr. PEALE at Philadelphia, which, with other specimens in that collection, is now probably lost for ever. Dr. HARLAN referred to it as *S. rufiventer*, but it did not agree with DESMAREST'S description of that species, as we ascertained by comparing it. On examining the description Dr. HARLAN gave of the specimen to which he referred, we ascertained that instead of describing it himself, he had, with slight variations, translated DESMAREST'S description.

PSEUDOSTOMA FLORIDANA.—Aud. and Bach.

Southern Pouched Rat.

PLATE CL. Fig. 1.—Old Male.

Unica per longitudinem stria in dentibus qui secant superioribus ; corpore P. bursario paullo exiguiore et minus robur prodente ; sacculis genarum minoribus ; palmis multo angustioribus ; caudâ longiore ; pilis crassioribus. Colore. supra subrufo-fusco, infra cinereo.

CHARACTERS.

A single longitudinal groove in the upper incisors ; body, rather smaller and less stout in form than P. bursarius *; cheek-pouches, smaller ; palms, much narrower ; tail, longer ; hair, coarser. Colour, above, brownish-yellow, beneath, gray.*

DESCRIPTION.

The body of this species is a little smaller and more slender and elongated than that of *P. bursarius ;* head, small ; nose, long, and not so blunt as in that animal. The fore-foot (or hand) has the palm narrow, and less tuberculated beneath than in *P. bursarius ;* nails, narrower, a little longer, and much less arched than in that species ; and the cheek-pouches are smaller. Tail, long (double the length of tail of *P. bursarius*), and has a little tuft of hair around the base ; the rest of it, however, is naked. Feet, naked, instead of clothed with hair as in that animal.

In the upper jaw the incisors are of moderate size, narrow, with a single groove in the centre, and no groove on the inner edges, as in *P. bursarius.*

Septum, naked, with the nostrils entering in at the sides, immediately at the roots of the incisors ; whiskers, rising from the sides of the nose, short, thin, and sparse ; eyes, small, placed near each other in the head ; the ears exhibit a slight margin around the auditory opening ; they are placed far back, and not distant from each other ; toes, five on each foot ; on the fore-feet the middle toe with the nail is much the longest ; the inner posterior toe is the smallest ; there are a few short rigid hairs on the inner edge of the palm, but the foot may be described as naked ; on the hind-foot, the claws, which are a little longer than those of *P. bursarius,* are hooked, and channelled beneath ; the hind-feet, to above the tarsus, are naked.

The hairs on the body are short, the coat not being half the length of the hairs of *P. bursarius*, and feeling much coarser and more rigid than in that species, especially on the under surface ; the cheek-pouches are somewhat differently situated from those of *P. bursarius :* whilst in the latter species the upper edge is more than half an inch below the base of the superior incisors, the cheek-pouches in the present species open immediately into the mouth, the upper edge reaching them, so that while in *P. bursarius* the food has to be taken from the pouches and conveyed round to the mouth, the present species is able, by the peculiar form and situation of the opening of its pouch, to shove the food from the pouch immediately into the mouth. The pouches are, internally, sparsely covered with short hairs.

COLOUR.

Hair on the back, plumbeous from the roots for three fourths of its length, then yellowish, tipped with black ; on the belly, the hairs are cinereous at base, and dirty yellow at the tips ; under the throat, they are of a uniform ashy white. Whiskers, white, with a few (shorter ones) dark brown ; teeth, pale orange ; claws, light yellow, those on the hind-feet dark brown at the points ; feet and tail, flesh colour. The result of the colouring of the hairs just mentioned is—back, brownish-yellow : nose and forehead, brown ; under surface from the chest to the thighs, bluish-gray ; throat, ashy white.

DIMENSIONS.

	Inches
From point of nose to root of tail, - - - - -	8¾
Tail, - - - - - - - - -	4
Longest middle claw, - - - - - - -	¾
Palm, including claw, - - - - - - -	1⅛
Breadth of head between the eyes, - - - - -	⅝
" between ears, - - - - - -	1⅛

HABITS.

The Southern Pouched Rat is very similar to the Canada Pouched Rat in its habits and manner of living, the chief differences in these respects between the former and the *Pseudostoma bursaria* being the natural result of different climate and situation.

This species is very remarkable for the apparently definite line of country it occupies, for, as far as we have been able to ascertain, although found in many places up to the southwestern bank of the Savannah river in Georgia, not one has ever been seen in South Carolina, or east of that river. This is the more singular as the wide range of the other species of this genus would lead us to suppose it not at all likely to be restricted by any fresh-water river, and indeed we can conceive no reason why it should not reach even to North Carolina and portions of Virginia, where sandy soils and dry pine lands similar to those it most frequents in Georgia, Florida, Alabama, and Mississippi, are widely extended.

Strangely enough, the common name applied to this animal where it is found is " Salamander."

The Southern Pouched Rat does not, like the *Pseudostoma bursarius*, remain under ground during the winter months, in a most probably dormant state, but continues its diggings throughout the year, and devours quantities of roots and grasses. It has hitherto been more frequently found living in the woods than near cultivated fields and plantations, but as the country becomes more settled will doubtless prove as great a pest in the gardens as its more northern relative, for an account of which see our first volume, pp. 332–339.

GEOGRAPHICAL DISTRIBUTION.

This species is found in the high pine barren regions, from the middle of Georgia and Alabama to the southern point of Florida, as far as the elevated portions of that State extend south.

We received two specimens from Major LOGAN in Dallas county, Alabama, several from Ebenezer, about twenty-five miles above Savannah in Georgia, and a number from the vicinity of Saint Augustine in East Florida.

We have not been able satisfactorily to ascertain its western range. We believe, however, it is not found west of the Mississippi. It is somewhat singular that this species is found on the very banks of the Savannah river, on the western side, and that notwithstanding, no traces of it have ever been seen east of that river, nor indeed in any portion of South Carolina, although there are extensive regions of high pine lands in that State which appear to be well suited to its habits.

GENERAL REMARKS.

It is highly probable that this is the species referred to by RAFFINESQUE and others as the Georgia Hamster ; inasmuch, however, as it was probably never seen by RAFFINESQUE, and he most likely formed his new genus

Geomys from figures representing the cheek-pouches as rising within the mouth, and hanging like sacs under the throat, we have thought it as well to decline adopting his genus thus founded in error, and to omit quoting him in any part of our work as an authority.

SOREX DEKAYI.—Bach.

PLATE CL. Fig. 2.—Young Male.

Magnitudine Arvicolæ Pennsylvanicæ ; colore supra ferrugineo, ex cinereo et flavo variegato, infra cinereo ; caudâ brevi atque cylindraceâ.

CHARACTERS.

Size of Arvicola Pennsylvanica ; rusty yellow gray colour above, cinereous beneath ; tail, short and cylindrical.

SYNONYMES.

Sorex DeKayi. Bachman, Jour. Acad. Nat. Sci., vol. vii., part 2, p. 377, pl. 23, fig. 4.
    "   "   DeKay, Nat. Hist. of New York, p. 17, pl. 5, fig. 2.

DESCRIPTION.

Dental Formula.—Incisive $\frac{2}{2}$; Canine $\frac{5-5}{2-2}$; Molar $\frac{4-4}{3-3} = 32$.

The two upper incisors are much curved, and pointed at tips ; the lateral incisors are each crowned with two tubercles except the fifth, which is smooth ; each grinder, on the upper surface, is furnished with four sharp points ; in the lower jaw the incisors are also much curved ; the first canine tooth is smaller than the second, and the molars are similar to those of the upper jaw. The body bears a resemblance to that of the shrew mole in shape. Head, rather short ; nose, distinctly bilobate ; nostrils, on the sides ; the eye is a mere speck ; there are no external ears ; whiskers, the length of the head ; the feet are more robust than those of any American Shrew we have examined, and are haired on the soles ; feet, clothed with short fine hairs ; the tail in the dry specimen is square, examined in the flesh is rounded, slightly dilated in the middle, and covered with short hair ; hind-foot, three middle nails nearly equal, outer toe a little longer than the inner, which latter is the shortest.

COLOUR.

Nose, feet, and nails, reddish-brown ; upper surface of body, rusty yellow gray ; a shade lighter on the under surface ; whiskers, for half the length from their roots cinereous, whitish at the tips ; incisors, black.

DIMENSIONS.

Female, captured in the garden at Minniesland near New York.

	Inches.
From point of nose to root of tail, - - - - -	4½
Tail (vertebræ), - - - - - - -	1
" (to end of hair), - - - - - -	1 1/10
From heel to point of longest nail, - - - - -	⅜
Breadth of fore-foot. - - - - - -	¼

Male, one eighth of an inch longer than the female.

HABITS.

We have always found it difficult to obtain satisfactory information as to the habits of the smaller quadrupeds, from the fact that many of our farmers and their men are unacquainted with the generic and even specific names, and consequently often mistake the habits of some genus or species for those of a very distinct one. The various species belonging to the genera *Scalops, Condylura,* and *Sorex,* are in most cases called (and considered to be) "ground moles," and thus are represented as all possessing the same habits. The gardener who caught for us the two specimens above described, said they were ground moles. On showing him that they were smaller, and had very different feet from those of any animal belonging to the genus *Scalops*, he said "they were only young ones, that their feet would become large by next year." On asking him about their nests, he said they were not old enough as yet to have young. When we requested him to show us their holes he first directed us to the ridges made in the soil by the ground mole.

After a careful examination, however, we ascertained that DEKAY's Shrew burrows deeper in the earth than *Scalops aquaticus*. The galleries of *S. DeKayi* run along at the depth of about a foot from the surface, and have apertures leading up to the open air at short distances from each other, by which the animals have ingress or egress. Ground moles seek worms and insects in the earth, whereas the Shrews come abroad on the surface, and run over the ground at night in quest of food, a habit in which the mole does not appear to indulge.

GEOGRAPHICAL DISTRIBUTION.

We received specimens of this small animal from Mr. COOPER, who obtained them in New Jersey; also one from Albany. We were present when two were captured near New York, and have neard of its existence in New England, Maryland, and Virginia.

GENERAL REMARKS.

We have seen specimens of DeKay's Shrew which exhibited a dark slaty gray appearance on the back and sides, and differed materially in colour from those from which we described.

This we attributed to the dark gray ones having been killed in the autumn or towards the approach of winter. Dr. DeKay seems to have described a specimen with this slaty coloured fur ; he gives its colour as " dark bluish throughout."

Plate CL

Nº 30

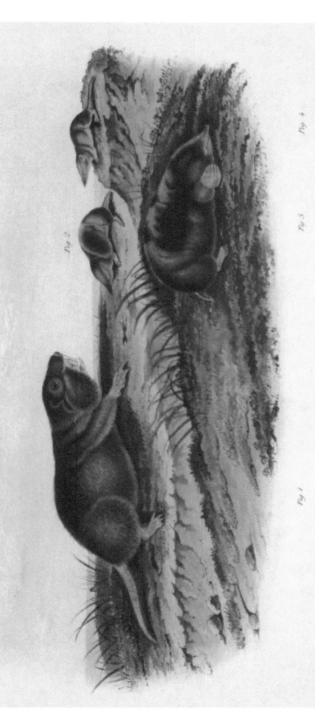

Fig 1

Fig 2

Fig 3

Fig 4

On Stone by Wm E Hitchcock

Fig.1 Southern Pouched Rat. Fig.2 Dekays Shrew. Fig.3 Long . Nosed Shrew. Fig.4 Silvery Shrew. Mole.

Drawn from Nature by J.J Audubon.

Lith: Printed & Col.d by J.T Bowen, Phil.

SOREX LONGIROSTRIS.—Bach.

LONG-NOSED SHREW.

PLATE CL. Fig. 3.—MALE.

S. Castaneus, rostro longo, caudâ longâ, auriculis amplis, vellere non occultis.

CHARACTERS.

*Nose, long ; **ears**, large, and prominent ; tail, long ; general colour, chesnut.*

SYNONYMES.

SOREX LONGIROSTRIS—LONG-NOSED SHREW. Bach., Jour. Acad. Nat. Sci., vol. vii., part 2, p. 370. Anno 1837.
OTISOREX PLATYRHINUS. DeKay, Nat. Hist. State of New York, p. 22, pl. 5, fig. 1. 1842.

DESCRIPTION.

Dental Formula.—Incisive $\frac{2}{2}$; Canine $\frac{5-5}{2-2}$: Molar $\frac{4-4}{3-3} = 32$.

Nose, very long ; whole upper jaw bordered with whiskers, extending to the middle of the ear ; lower jaw, sparsely covered with the same kind of hair, but shorter ; extremity of the muzzle, naked, deeply indented and bilobed ; the eyes are distinctly visible, and larger than in most species of this genus ; the ear extends considerably beyond the fur, is comparatively large and thickly clothed within and without, with short soft hairs ; the auditory opening is covered with a large oblong lobe on which are sprinkled a few stiff long hairs ; tail, nearly round, but in the dried specimen becoming square ; it is clothed with short hair above and beneath, as also the feet and palms to the extremities of the nails ; toes, five ; the whole body is slender, and the feet small and weak. The fur is close, very fine, and glossy.

COLOUR.

Above, uniform chesnut ; beneath, a shade lighter ; points of the teeth, dark brown ; nails, horn colour, tipped with black.

VOL. III.—32

DIMENSIONS.

		Inches.
Length from the nose to the origin of the tail,	- -	$1\frac{7}{8}$
" of tail, - - - - - - -	-	1
" of head, - - - - - - -	-	$\frac{7}{8}$
Height of ear, - - - - - - - -	-	$\frac{1}{4}$
Length of hind-foot from heel to end of nails, -	-	$\frac{3}{8}$

We have since measured a specimen procured at Tulula falls in Georgia, the dimensions of which were as follows :

	Inches.	Lines.
Length from nose to origin of tail, - - -	2	2
" of tail, - - - - - - -	1	3

HABITS.

We possess very little knowledge in regard to the habits of this little Shrew. The first specimen we saw was obtained in the swamps that border the Santee River, by Dr. ALEXANDER HUME; his labourers found it whilst digging a ditch through grounds nearly overflowed with water. Another was obtained in a singular manner. Whilst we were at the house of Major LEE in Colleton district, his huntsman brought in some wild ducks, and among the rest a hooded merganser (*Mergus cucullatus*). There was a protuberance on the throat of this bird, appearing as if it had not fully swallowed some article of food at the time it was shot. On opening the throat, it was found to contain this little Shrew, which was fresh, and not in the least mutilated.

We saw two or three Shrews in the same vicinity which we think were of this species, coming out of a bank on the edge of a rice field and swimming in the canal at the dusk of the evening. From the above circumstances we are induced to think that this quadruped prefers low swampy situations, and is to a certain extent aquatic in its habits. We more recently obtained a specimen from Col. W. E. HASKELL, of St. Paul's parish, South Carolina, which was a shade lighter, and a little larger, than others in our possession, but presented no specific differences.

GEOGRAPHICAL DISTRIBUTION.

The Long-nosed Shrew (although apparently very sparingly) is found in South Carolina. We saw a specimen in the possession of the keeper of the public house at the Tulula falls in the mountains of Georgia. It evidently extends throughout the middle States, and has been taken in New York and New England.

GENERAL REMARKS.

The American Shrews may be easily arranged into three natural groups ; first, those with short ears and tail, of which *Sorex DeKayi* would form the type ; second, those with large palms, broadly fringed, such as *S. fimbripes* ; and third, those with long ears and tail, of which the present species would be the type.

We perceive that Dr. DeKay has formed the present species into a new genus, *Otisorex ;* but as the European naturalists had previously proposed a number of genera, one of which would include the present species, we prefer for the present leaving our American Shrews in the genus *Sorex*. We have no hesitation in saying that Dr. DeKay's species *platyrhinus*, on which his genus was founded, is identical with *S. longirostris*, the subject of our present article.

SCALOPS ARGENTATUS.—Aud. and Bach.

SILVERY SHREW-MOLE.

PLATE CL. Fig. 4.—FEMALE.

S. Pilis tota longitudine albo, plumboque annulatis; fronte, mentique albido flavescente.

CHARACTERS.

Hairs, from the roots regularly annulated with white and plumbeous; fore-head and chin, yellowish-white.   Colour of the body, shining silver gray.

SYNONYME.

Scalops Argentatus—Silvery Shrew-Mole. Aud. and Bach., Jour. Acad. Nat. Sci., October 5, 1841.

DESCRIPTION.

In form this species is cylindrical, like the common Shrew-Mole (*S. aquaticus*), to which it bears a strong resemblance.   Muzzle, naked; and the nostrils inserted, not on the sides, as is the case in *Scalops Breweri*, but in the upper surface, near the point of the nose, as in *S. aquaticus*.   Eyes, not visible, and appear covered by an integument; the lips are fringed with rather coarse hairs; this species is pendactylous, with naked palms and tail; the teeth are larger, shorter, and broader than those of the common Shrew-Mole; the fur is long and lustrous on the back, but much shorter and more compact on the under surface.

COLOUR.

Teeth, and nails, white; palms, hind-feet, and tail, flesh coloured; nose, forehead, lips, and chin, yellowish-white; the fur on the back is from the roots marked with alternate narrow bars of dark blue and white to near the extremities, where it is broadly barred with ashy white, and so slightly tipped with brown that the lighter colour beneath is still visible on the surface, giving it a beautiful silvery appearance, which presents a variety of changes, on being exposed to different rays of light.   On the lower surface the hair is plumbeous from the roots to near the tips, where it is barred with whitish; it is tipped with light brown. There is a spot of white on the centre of the abdomen, which is apparently accidental, as we have occasionally observed it in other species of this genus, as well as in the true mole (*Talpa*) of Europe.

DIMENSIONS.

	Inches.	Lines.
Length of head and body, - - - - - - -	7	1
" tail, - - - - - - - -	1	
Breadth of palm, - - - - - - -		10
From tarsus to point of longest nail, - - - -		7

HABITS.

Dr. GEO. C. LEIB, who discovered this animal in the prairies of Michigan, gave us no account of its habits, which we presume are similar to those of the common Shrew-Mole.

GEOGRAPHICAL DISTRIBUTION.

We have not heard of this beautifully furred Mole in any other locality than that where our specimen was procured, which is the only one we have ever seen, and the one from which our figure and description have been made.

GENERAL REMARKS.

Of the several species of Shrew-Mole that inhabit North America, this in point of colour is the most brilliant that has yet been brought to the notice of naturalists. Although it bears a general resemblance to the common Shrew-Mole, yet the characters it presents have induced us after some hesitation and doubt, to designate it as a new species. It is nearly double the size of the common Shrew-Mole; the fur is much longer and softer, and differs strikingly in colour and lustre. Our specimen was evidently a young animal, although the dentition was similar to that of *Sc. aquaticus*. Some of the small thread-like teeth that are placed behind the incisors in the upper jaw were wanting on one side, and were only barely visible on the other. The young of *Scalops aquaticus* have but thirty teeth until they are more than a year old; when they have arrived at their full vigour they are furnished permanently with thirty-six. The skulls of *Scalops Townsendii* and *S. Brewerii* each contain forty-four teeth.

Before we take leave of the Shrew-Moles of our country, we have to add that RICHARDSON (F. B. A. p. 12), in noticing the assertion of BARTRAM, that a true mole, *Talpa*, existed in America (in which he was supposed by later writers to be mistaken), asserts that there are several true moles in the museum of the Zoological Society of London, which were brought from America, and which differ from the ordinary European species (*Talpa Europea*), in being of a smaller size and having a shorter and thicker snout, their fur being brownish-black. DEKAY, in the Natural History of New York (p. 16), refers to the above statement. We however examined these specimens in the Zoological Museum, and found they consisted of only two species—our common *Scalops aquaticus*, which RICHARDSON strangely mistook for another species, and *Scalops Breweri*, to which he particularly referred. Thus far therefore no true specimen of the genus *Talpa* has been discovered in America, and we have no doubt that the species referred to by BARTRAM as the black mole was BREWER'S Shrew-Mole, which in certain lights appears quite black.

VULPES UTAH.—Aud. and Bach.

JACKALL FOX.

PLATE CLI.

V. corpore grandiore, pilis velleris longioribus nec non gracilioribus quam in V. fulvo, cauda magna cylindracea.

CHARACTERS

Larger than Vulpes fulvus ; fur longer and finer than in that species ; tail large and cylindrical.

SYNONYMES.

VULPES UTAH.—Aud. and Bach., Proc. Acad. Nat. Sci., Phil., 1852, p. 114.
VULPES MACROURUS, Baird, Stansbury's Report.

DESCRIPTION.

Claws slightly arched, compressed, channelled beneath, horn color ; hair, of two kinds, first, a coarse and long hair covering the fur beneath it ; second, a dense and very soft fine fur, composed of hairs that are straight, but crimped and wavy, as in the silver grey fox. Fur plumbeous at the roots, gradually becoming dark brown towards the tips in those parts of the body which are dark colored on the surface ; in those parts which are white, the fur is white from the roots, and on no part of the animal does it present any annulations.

The long hairs are dark-brown from the roots, yellowish-white near the middle of their length, and are tipped with black.

On the under surface the hairs are principally white their whole extent, with a few black ones intermixed ; the fur on the tail is rather less fine and more woolly than on the body.

Feet covered with soft hair reaching beyond the toes ; on the forehead the hair is rather coarse and short, with fine fur beneath.

COLOUR.

Greyish-white on the head, dark brown on the neck, greyish-brown on the dorsal line and on the sides ; the throat, under surface of the body, insides of legs, and feet black.

The tail is irregularly banded with dark brown and dull white, the tip white for about three inches.

Another Specimen.—Nose, both surfaces of the legs, and behind the ears, dark, reddish-brown ; whiskers black ; under side of neck, and a line on the belly, liver brown. Fur on the back very fine, and dark ashy-grey from the roots : the longer hairs on the back are black at the roots, and are broadly tipped with white ; fur on the sides, cinereous at the roots, and yellowish-white from thence to the end.

There is a reddish tinge on the neck, extending to the shoulders ; sides of the face grizzly-brown ; the hair on the tail is irregularly clouded with brown and dull white, and is lightest on the under surface.

DIMENSIONS.

	Feet.	Inches.
From point of nose to root of tail,	2	8
Tail, (vertebræ,)	1	4
" (to end of hair,)	1	8
Circumference of tail, (broadest part,)	1	8
From shoulder to fore-feet,	1	5
From rump to hind-feet,	1	6
Height of ears, (posteriorly,)		4
From point of nose to eye,		3⅛
Longest hairs on the brush,		5
" on the body,		3

HABITS.

This animal was first noticed, by LEWIS and CLARK, as the large Red Fox of the plains, (vol. 2, p. 168,) and was referred to by us in the first volume of the Quadrupeds of North America, p. 54, where we described it from a hunter's skin.

Having obtained a beautiful specimen from Captain RHETT, of the United States Army, we gave it the name of *Vulpes Utah*, as it is, so far as our information extends, chiefly found in the Utah territory, although it probably ranges considerably north of the Great Salt Lake.

The habits of this beautiful Fox are similar to those of the Red Fox, and it runs into many varieties of color.

Captain RHETT informed us that he killed the specimen, kindly presented to us by him, near Fort Laramie.

Several specimens of *Vulpes Utah* have been received at the Smithsonian Institution, and it will probably soon be well known.

GEOGRAPHICAL DISTRIBUTION.

This Fox, as we have ascertained since writing the above, is procured

Drawn from Nature by J.W. Audubon.

On Stone by Wm E. Hitchcock

Lith. Printed & Col.d by J.T. Bowen, Phil.

Jackall Fox.

throughout the Rocky Mountain regions, although by no means abundantly, as far north as the traders of the fur companies push their outposts. It is found also in Oregon.

GENERAL REMARKS.

The exploring expedition sent by the United States, (1838 to 1842) did not procure any specimens of this Fox, although we find by Mr. PEALE's Catalogue, they obtained the Vulpes Virginianus, in both Oregon and California.

SCIURUS MUSTELINUS.—Aud. and Bach.

Weasel-like Squirrel.

PLATE CLII.—Male. Fig. 1.

S. Cervice longissima; caudâ corpore longiore; pilis curtis, rigidis, compressis, teretibus; omni corporis parte nigerrima.

CHARACTERS.

Neck, very long; tail, longer than the body; hair, short, rigid, adpressed, and glossy; the whole body, jet black.

SYNONYME.

Sciurus Mustelinus—Weasel Squirrel. Aud. and Bach., Proc. Acad. Nat. Sci. Phil., Oct. 5, 1841, p. 32.

DESCRIPTION.

The unusually long neck of this species, together with its long slender body, and smooth lustrous hair, give it somewhat the appearance of a weasel, and suggested to us the specific name.

Ears, of moderate size, and nearly naked, there being only a few hairs on the borders; feet, covered with very short hairs, which only reach to the roots of the nails; tail, long, not bushy, moderately distichous.

COLOUR.

The hairs, in every part of the body, are deep black from the roots to the tips, and the surface is glossy.

DIMENSIONS.

	Inches.	Lines.
Length of head and body - - - - - -	10	
" tail, - - - - - - -	13	
From shoulder to point of nose, - - - -	3	10
Tarsus, - - - - - - - -	2	5
Height of ear posteriorly, - - - - -		6

HABITS.

The Weasel-like Squirrel feeds in the woody portions of California, on acorns, the seeds of the pines and other trees, and makes its nest in the oaks or nut-bearing pines of that country, which, from their broad spreading branches and dense leafy boughs, afford it security against the hunter, as with equal cunning and agility it hides itself, when alarmed, amid the evergreen foliage, and except when surprised on the ground or near the earth, and shot instantly, can seldom be killed. There is no more tantalizing game, in fact, and as the branches interlock at a moderate elevation from the ground, the animal easily goes from one tree to another, and so swiftly that it is not often to be traced in its course of flight along the boughs.

We are unacquainted with the time of this animal's breeding, but presume it brings forth about four or five young at a birth. The young of all species of squirrels with which we are familiar, are born blind, and remain without sight from four to six weeks. This is an admirable provision of nature for their safety, as were they able to use their eyes at an earlier period, they would doubtless be tempted to quit the security of the nest and venture on to the branches, before they had gained strength enough to preserve their footing, and would thus probably fall to the earth and be killed.

GEOGRAPHICAL DISTRIBUTION.

The specimen from which our figure and description were made was procured in California. We have no authority for stating its northern or southern range, but consider it a western species—by which we mean that it is not found east of the Rocky Mountain chain.

GENERAL REMARKS.

From its thin covering of hair, being nearly destitute of the soft fur usually clothing the squirrels, this species may be considered as belonging to a moderate or warm climate. It differs widely from all the other species of Black Squirrel (as well as all black varieties of Squirrel), in our country. It has shorter and coarser hair than *S. capistratus*, and is destitute of the white nose and ears of that species, with none of the white tufts invariably found in *S. niger;* and has a smaller body, although a much longer tail than *S. Auduboni*, without the white, yellow, and brown annulations in the hair which characterize that species.

SCIURUS AUDUBONI.—Bach.

LARGE LOUISIANA BLACK SQUIRREL.

PLATE CLII. Fig. 2.—MALE.

Paulo minor quam Sciurus Niger; aures breviores; dentes qui cibum secant latiores; cauda longitudine corpori par; capilli valde crassi, tactuque asperi, sed nihilominus nitidi. Color, supra niger; infra subfuscus.

CHARACTERS.

A little less than Sciurus niger; *ears, shorter; incisors, broader; tail, as long as the body; fur, very coarse and harsh to the touch, but glossy; colour, above, black, beneath, brownish.*

SYNONYME.

LARGE LOUISIANA BLACK SQUIRREL—SCIURUS AUDUBONI. Bach., Monog. of the Genus Sciurus, p. 33. 1839.

DESCRIPTION.

Dental Formula.—Incisive $\frac{2}{2}$; Canine $\frac{0-0}{0-0}$; Molar $\frac{4-4}{4-4} = 20$.

Our specimen has the above number of teeth. If the small anterior molar in the upper jaw exists in the young, which we suspect to be the case in all American species, it is deciduous; and we are warranted in arranging this species among those which have permanently but twenty teeth. In the upper jaw the anterior molar is triangular in shape, and crowned with three blunt tubercles; the other molars are quadrangular, with concave crowns.

Head, narrower, and body, thinner than in *S. niger;* ears, short and conical, covered on both surfaces with short adpressed hairs, presenting no tufts; whiskers, longer than the head, extending to the shoulders. Fur on the back, very coarse.

COLOUR.

Incisors, deep orange; whiskers, black; back, upper parts, outsides of limbs, and feet, black, with a faint tinge of brown. Many of the hairs are

Fig 1

Fig 2

Drawn on Stone by Wm. E. Hitchcock

Fig 1 Weasel like Squirrel

Fig 2 Large Louisiana Black Squirrel.

Drawn from Nature by J.W. Audubon

Lith. Printed & Col.d by J.T.Bowen, Phil

however obscurely annulated with yellowish-white. The whole under surface, and the inner sides of the legs, are brownish. Most of the hairs on the under surface are grayish-white at the base, some are annulated with black and yellow, and others are brown.

Chin, black, with the extreme tip whitish ; end of nose, brownish ; tail, black ; when viewed beneath, the hairs exhibit deep yellow annulations ; most of the hairs are brownish towards the tip.

DIMENSIONS.

	Inches.	Lines.
Length of head and body, - - - - -	11	6
" tail (vertebræ), - - - - -	8	9
" " (to end of hair), - - - - -	11	6
" palm to end of middle fore-claw, - -	1	6
Heel to point of longest nail, - - - - -	2	6
Height of ear posteriorly, - - - - -		3
Length of fur on the back, - - - - -		6

HABITS.

This southern Black Squirrel was first described by Dr. BACHMAN, from a specimen obtained by J. W. AUDUBON in Louisiana, and was named by him after its discoverer. It frequents high grounds, and has all the active, restless, and playful habits of the genus.

GEOGRAPHICAL DISTRIBUTION.

The Louisiana Black Squirrel has been seen west of the Mississippi, and as we think is occasionally found in Texas. It is sometimes offered for sale in the New Orleans markets, being shot in the neighbourhood of that city.

GENERAL REMARKS.

We have been informed by some officers of the United States army that a Squirrel similar to the present species is found in Texas and in parts of New Mexico, but from there being no specimens we could not positively identify the Black Squirrels these gentlemen had observed with *S. Audubont.*

SCIURUS ABERTI.—Woodhouse.

Col. Abert's Squirrel.

PLATE CLIII.—Fig. 1.

S. Auribus magnis latisque, cristatis longis subnigris cinereisque crinibus ; rubra in dorso striga.

CHARACTERS.

Ears large and broad, tufted with long blackish grey hairs ; a reddish stripe on the back.

SYNONYMES.

Sciurus dorsalis.—Woodhouse, Proc. Acad. Nat. Sci., Phil., June, 1852, p. 110.
Sciurus Aberti.—Woodhouse, Proc. Acad. Nat. Sci. Phil., Dec., 1852, p. 220.

DESCRIPTION.

Ears large and broad, with very long tufts ; tail very large ; fur long, compact, and soft ; claws long, very strong, and much curved ; whiskers very long.

COLOUR.

General colour above dark grey, with the exception of the dorsal line and a band extending along the external base or hind part of the ear, which is of a rich ferruginous brown colour ; beneath, white, with the exception of the perineum, which is grey ; cheeks greyish white ; tail grey above with a broad white margin, and white beneath ; claws of a black colour with the exception of their points, which are light and almost transparent ; whiskers black ; iris dark brown.

DIMENSIONS.

Dried Skin.	Inches.
Length from nose to root of tail, about, - - - -	13
From heel to point of longest nail, - - - - -	$2\frac{3}{10}$
Height of ears, externally, - - - - -	$1\frac{3}{10}$
" " to end of tufts, - - - -	$2\frac{3}{10}$
Breadth " ————, - - - - -	1
From ear to point of nose, about - - - - -	$1\frac{7}{10}$
Tail (vertebræ), about - - - - - - -	8
" to end of fur, - - - - - - - -	11

HABITS.

Dr. WOODHOUSE, from whose description we have extracted above, makes the following remarks: "This beautiful squirrel I procured whilst attached to the expedition under command of Capt. L. SITGREAVES, Topographical Engineer U. S. Army, exploring the Zuni and the great and little Colorado rivers of the west, in the month of October, 1851, in the San Francisco Mountain, New Mexico, where I found it quite abundant, after leaving which, I did not see it again."

GEOGRAPHICAL DISTRIBUTION.

So far as shown by the foregoing account, and according to our knowledge, this squirrel has not been seen except in the San Francisco Mountain, New Mexico. It is, however, most likely that it inhabits a considerable district of elevated and wooded country in that part of our Continent, and may hereafter be found in California or even Oregon.

GENERAL REMARKS.

We have not been able to procure any further information regarding this species, which was first named *Sciurus dorsalis* by its discoverer, but a subsequent examination having satisfied him that this name had "already been applied by J. E. GRAY, to one of the same genus," he proposed "to call it *Sciurus Aberti*, after Col. J. J. ABERT, chief of the corps of Topographical Engineers, U. S. Army, to whose exertions science is much indebted."—(Proceed. Acad. Nat. Sci., Phil., Dec. 1852, p. 220.

It gives us great pleasure to welcome this beautiful new animal under the name of Col. ABERT's Squirrel.

SCIURUS FOSSOR.—Peale.

PLATE CLIII.—Fig. 2.

S. Supra e nigro alboque intermixtis griseus, subtus albus, auribus magnis, breviter pilosis, naso nigro, cauda disticha, albo-marginata, corpore longiore.

CHARACTERS.

Above, grey ; beneath, white ; ears not tufted, but clothed within and without with short hairs ; nose black ; tail distichous, tipped with white ; body long and rather slender.

SYNONYMES.

Sciurus Fossor.—Peale, Mam., &c., of the U. S. Exp. Exped., 1838–42.   Phila. 1848.
Sciurus Heermanni.—Dr. Le Conte, Proceed. Acad. Nat. Sci., Phil., Sept., 1852, p. 149.

DESCRIPTION.

Whiskers shorter than the head ; ears large, subtriangular, rounded at the tip, and covered both within and without with short hair, which does not in any way form a fringe at the margin ; tail long and distichous, with long hairs which are grey at the roots, black above and tipped with white ; body long and rather slender ; hair on the body long and not fine.

COLOUR.

Body above light grey, produced by an intermixture of black and white points ; the hairs are grey at base, then black, and have a pure white annulation about the middle ; intermixed with them are a few longer pure black hairs. A small spot towards the tip of the nose, and an indistinct line above the eyes are black ; whiskers black. Beneath, the body is pure white, except the perineum, which is grey ; tail grey, blackish towards the edges, and broadly margined with white ; rather lighter in colour beneath.

DIMENSIONS.

	Inches.	Lines.
From tip of nose to root of tail, - - - -	12	5
Head, - - - - - - - -	3	2
Length of ear, - - - - - - -		9
Breadth of ear, - - - - - -		7
Fore foot to end of longest claw, - - - -	2	1
Hind foot to end of longest claw, - - - -	3	2
Tail, to end of vertebræ, - - - - - -	9	8
hair, - - - - - -	13	

Plate CLIII

Fig. 1

Fig 1 Scol. Abert's Squirrel. — Fig 2. California Grey Squirrel.

Drawn from Nature by J W Audubon

Lith Printed & Cold by J T Bowen, Phil

HABITS.

This beautiful squirrel has been often killed by Mr. J. E. CLEMENTS, in the pine woods of California, near Murphy's "diggings." It is exceedingly swift on the ground, and will not readily take to a tree, or, if it does, ascends only a few feet, and then jumping down to the ground runs off with its tail held up but curved downwards towards the tip like that of a fox when in flight.

By the aid of a fast cur dog, it may, however, be put up a tree. In this case it hides if a hole offers in which to conceal itself; and unlike some others of its genus, seldom leaps from one tree to another over the higher boughs in the endeavour to make its escape.

It appears to make its nest generally in the decayed part of an oak tree, and in the desire to reach its secure retreat, is doubtless led to attempt to run to this tree on the ground, rather than by ascending the nearest trunk and jumping from branch to branch.

A large part of its food consists of nuts, which are stuck in hollows or holes bored in the pine trees by a species of woodpecker called by the Californians "Sapsuckers." These nuts are placed in holes in the bark, which are only so deep as to admit the nuts (which are placed small end foremost in them), leaving the large end visible and about flush with the bark—they thus present the appearance of pins or pegs of wood stuck into the trees, and are very curious objects to the eye of the stranger.

The California grey squirrel is a roving animal. One may sometimes see from one to a dozen in a morning's hunt in the pines, and again not meet any. They very seldom leave the pines, but are occasionally seen in the dry season following the beds of the then almost empty water courses, which afford them, in common with other animals and birds, water and such roots and grasses as they cannot find on the uplands at that period of the year.

They bark somewhat in the same tones as the grey squirrel of the Atlantic States, but immediately cease when they perceive they are observed by man. Sometimes they seem to be excited to the utterance of their cries by the whistling of the California partridges, which, near the hills, approach the edges of the pine woods.

Most of those shot by Mr. CLEMENTS were killed when running on the ground.

GEOGRAPHICAL DISTRIBUTION.

This species is found in California in the wooded districts on the sides of the hills, and extends to Oregon, as, in Mr. PEALE's work, we have accounts of its having been observed there.

It is also almost a sure conclusion that it is found on the ridges of the mountains, as far south as the nut-bearing trees invite it, and it may thus reach quite a low latitude.

GENERAL REMARKS.

The pine nuts referred to in the account of the habits of this squirrel, as a favourite article of food for it, are placed on the cones of the *Sugar Pine*, (*Pinus Lambertii*, Douglas), so called from the gum which exudes from it, where the bark has been wounded, becomes hard and white, and is quite sweet to the taste.

The nuts are formed on the cones, sometimes twenty or thirty on one cone. The Indians pound and crack them. They are very good eating, and taste not unlike a hickory nut. The shell is thin, but hard, the nut covered with a skin like the peach kernel, &c.

We hesitated somewhat as to adopting the name (*Sciurus Fossor*) given to this species by Mr. TITIAN R. PEALE, as his volume on the " Mammalia and Ornithology " of the United States Exploring Expedition, &c., has been suppressed ; but as about one hundred copies, it appears, were circulated, we think it is only justice to Mr. PEALE to quote his work, which, as it was printed in 1848, gives his name the priority over *Sciurus Heermanni*, under which this species was described by our friend Dr. LE CONTE, September, 1852.

Its flesh is good eating, and it is sufficiently abundant in some parts of California to make it worth the hunting for market.

SPERMOPHILUS HARRISII.—Aud. and Bach.

Harris's Spermophile, or Marmot Squirrel.

PLATE CLIV.—Fig. 1.

S. Magnitudine *Tamiæ Lysteri ;* strigis duobus albis dorsalibus ab humeris ad femora ; fronte rufo-canescente, colle et ventre cinereis-albis, dorso fulvo-cinereo ; cauda disticha.

CHARACTERS.

Size of Tamias Lysteri ; a narrow white stripe on each side of the back, from the shoulder to the thighs ; forehead reddish-grey ; neck ashy-white running into yellowish iron-grey on the back ; under surface ashy white ; tail distichous.

DESCRIPTION.

Head small and delicate ; neck, rather long ; body slender ; legs rather long, a few long hairs growing out and fringing the hind parts of the fore legs ; the cheek pouches appear rather small in the dried specimen ; ears thinly clothed on both surfaces with short adpressed hairs ; short, somewhat triangular, and rather acute at the tips ; tail of moderate length. depressed at base, with the hairs growing from the side, giving it a decidedly distichous appearance ; the teeth resemble those of *T. Lysteri ;* are rather small, and the lower incisors are slightly curved.

Whiskers not numerous, reaching the ear ; hairs on the back very short, somewhat coarse, but lying very smoothly, giving the animal a glossy appearance ; on the under surface they are coarse and rigid. There are five toes on each foot ; on the fore-feet a small tubercle in place of a thumb, with a blunt nail ; second nail from the thumb longest, as in the rest of the spermophili ; on the hind feet the middle toes are longest, the two on each side being of nearly equal length, the outer considerably shorter, and the inner shortest ; claws slightly compressed, and a little curved ; feet clothed with short hairs, but which do not conceal the nails ; the eyes are of moderate size, and are placed midway between the point of the nose and the root of the ear ; soles of feet tuberculated and naked, except a few hairs between the toes.

COLOUR.

Incisors dingy yellow; whiskers and nails black; back and sides minutely speckled with white, on a yellowish-brown ground; the hairs are dark-brown at the roots, then white, then black, and the tips brownish-white, with a tinge of yellow; on the nose and forehead, the speckled appearance of the back is superseded by a rufous tint; between the ears, on the neck, and a little downwards, towards the legs, greyish-white is the prevailing color; a narrow white stripe, rising from behind the shoulder, and running along the side of the back to the middle of the hips, there loses itself in the general colour of the body; around the eye, throat, chin, inner surface of legs, and whole under surface of body, whitish, with a few black hairs interspersed; a tinge of brownish-red on the outer surface of the fore legs is more strongly red on the thighs; feet and outer surface of legs yellowish white.

The hairs of the tail are whitish at the roots, twice annulated with black, and tipped with white.

There is a line of whitish yellow on the flanks, separating the colour of the back and sides from the under surface distinctly, and extending along beneath the reddish brown tint on the thighs, where it becomes a deeper yellow.

DIMENSIONS.

	Inches.
Length of head and body - - - - - - -	5¾
" tail (vertebræ) - - - - - - -	3¼
" " to end of hair - - - - - -	4½
From tarsus to end of longest claw - - - - -	1½
Length of fore leg from the shoulder - - - -	2
" hind leg from the thighs - - - - -	2½
Breadth of tail, when distichously arranged - - -	1¾
Height of ear (posteriorly) - - - - - - -	⅜
Longest claw on the fore foot - - - - - - -	⅛

HABITS.

There is nothing to be said by us about the habits of this species, as it has not been observed, so far as we know, since our specimen was procured, and we have not even a knowledge of the precise locality in which it was obtained by Mr. J. K. TOWNSEND, who gave the specimen to our esteemed friend, EDWARD HARRIS, Esq., from whom we received it some time since, and with whose name we have honoured this pretty little animal.

Fig 1

Fig 2

Fig 1. Harris's Marmot Squirrel.   Fig 2. Californian Meadow Mouse.

GEOGRAPHICAL DISTRIBUTION.

Probably west of the Rocky Mountains, on the route followed by Messrs. NUTTALL and TOWNSEND, in their journey to Oregon overland.

GENERAL REMARKS.

This species bears a very slight resemblance to *Spermophilus lateralis* of SAY, but differs so widely from it that it is unnecessary to institute a close comparison. It is a smaller animal, the head and ears being diminutive compared with the latter ; it has a single stripe of white on the sides of the back, whilst in *Lateralis* a broad white stripe is margined on each side by a stripe of black, giving it the appearance of having four black stripes on the back, while *S. Harrisii* has no black about the back or sides at all.

ARVICOLA EDAX.—Le Conte.

CALIFORNIA MEADOW MOUSE.

PLATE CLIV.—FIG. 2.

A. Brevis et robustus, supra spadiceo et nigro permixtus. Auribus extra pilos extantibus. Cauda mediocri, supra nigra, subtus cinerea.

CHARACTERS.

Body short and thick ; above, brown mixed with black ; ears not concealed by the hair ; tail moderate length, black above, beneath grey.

SYNONYME.

ARVICOLA EDAX.—Le Conte, Proc. Acad. Nat. Sci., Phil., Oct. 1853, p. 405.

DESCRIPTION.

Head short and blunt ; ears round, not entirely concealed under the fur, hairy within and without, antitragus large and semicircular ; feet covered with short, shining grey hair ; thumb tubercle, with a short, very blunt nail ; tail of moderate length, hairy above.

COLOUR.

Hair plumbeous black above and on the sides, tipped with shining brown mixed with black ; beneath tipped with grey ; feet grey ; tail dusky above, grey beneath, with a slight brownish tinge.

DIMENSIONS.

	Inches.
Length (including the tail)	5.5
" of head	1.4
" " ears	.5
" " fore leg	1.3
" " hind leg	1.5
" " tail	1.5

HABITS.

This new Arvicola from California has doubtless the habits of the genus to which it belongs. Our friend Major LE CONTE does not, however, say anything in relation to its peculiarities, and as all our information of the existence even of the animal is derived from that gentleman, we have nothing to say further. We present our thanks to Major LE CONTE for the loan of the skin from which our figure was made.

GEOGRAPHICAL DISTRIBUTION.

California is named as the habitat of this Meadow Mouse; but we are not informed whether it is widely diffused there, or is confined to certain localities.

GENERAL REMARKS.

The description and dimensions of the California Meadow Mouse above given are quoted with slight alterations from Major LE CONTE's paper cited above.

PROCYON CANCRIVORUS.—Cuv.

CRAB-EATING RACCOON.

PLATE CLV.

P. Supra canescens plus minus in nigrum vergens, subtus flavo-albente, pedibus fuscescentibus, facie albidâ, fascia oculum circumcingente et cum oppositâ confluente nigra ; caudâ rufescente, annulis nigris.

CHARACTERS.

Body, above greyish, more or less shaded with black ; beneath, light yellow ; feet brownish yellow ; face whitish ; a black band surrounding the eye uniting with the opposite one ; tail reddish, annulated with black.

SYNONYMES.

URSUS CANCRIVORUS.—Cuv. Regne An., i., p. 138.
RATON CRABIER.—Buff. His. Nat., Suppl. vi., p. 236, t. 32.
AGUARA-POPÉ.—D'Azara, Essai i., p. 327.
PROCYON CANCRIVORUS.—Desm. in Nouv. Dict. xxxix., p. 93. 2.
"   "   Briggins, Paraguay, p. 213.
"   "   Prince Max. Wied, Beitrage ii., p. 301.
"   "   Griffith An. Kingd., Synopsis, Species 325, p. 114.
"   "   Weigmann, Arch. iii., p. 371.
"   "   Rengger, Paraguay, p. 113.

DESCRIPTION.

Body longer and more slender than that of the common Raccoon (*P. lotor*), legs longer, ears shorter, less rounded, and more pointed, and tail thinner than in the latter species. The tail diminishes towards the end. Hairs coarse ; nails prominent ; feet closely haired ; under-fur short and sparse.

COLOUR.

Point of nose black ; whiskers white and black, a blackish band around the eyes, extending nearly to the ears ; sides of the face, and above the eyes, and a spot on the forehead, whitish ; extremities of ears yellowish white, their bases dark brown ; nails black ; tail barred with black and white ; cheeks, jaws, under-part of the neck, breast, and belly, white, with a tinge of yellowish brown. Upper surface of body ash-brown.

Plate CIX

Fruit eating Raccoon

On Stone by Wm E. Hitchcock.

Lith. Printed & Col'd by J. T. Bowen, Phil.

Drawn from Nature by J.W. Audubon

DIMENSIONS.

						Inches.
From point of nose to root of tail,	-	-	-	-	-	22
Tail (vertebræ),	-	-	-	-	-	9
Point of nose to ear, -	-	-	-	-	-	4½
Fore leg to point of longest nail,	-	-	-	-	-	8
Thigh to point of longest nail, -	-	-	-	-	-	8
Breadth of skull,	-	-	-	-	-	3½

HABITS.

This Raccoon, as observed (in California) by Mr. J. E. CLEMENTS, generally conceals itself during the day in the oak trees which, from decay, afford holes into which it can retreat. It climbs with great agility up the rough bark of these until it reaches some decayed branch in which a cavity sufficiently large to hide in is found. There is a singular fact in this connexion, which is that most part of the rotted holes or places in these California oaks are found in the branches, not in the trunk. We are informed that many trees cut down for the purpose of making fence-rails, &c., are quite sound in the main stem, but the reverse in the branches, and that occasionally a large lateral branch will break down and fall to the ground—perchance startling the hunter who may be listening in hopes of hearing the sound of an approaching animal.

The food of this species consists of acorns, grapes, berries, eggs, birds, &c., and of late it has been known to attack chickens on the farms of the isolated settlers, sometimes endeavouring to take them off the trees adjoining the houses.

The flesh of these animals, when boiled first, and afterwards roasted, is very palatable, and not much unlike fresh pork. They are, however, generally lean, and by no means as fat as the Raccoon of our Atlantic States.

This species has been seen by Mr. CLEMENTS on more than one occasion, apparently keeping company with the black-tailed deer (*C. Richardsonii*), being on the mountains, following the same route, among several of these animals.

Two of those killed by Mr. CLEMENTS had been put up a tree by a dog during the night, and were discovered by the barking of the latter in the morning. They were only about half a mile from the house, and when approached, did not offer to come down, or otherwise attempt to escape. They had not ascended the tree more than some twenty feet from the ground.

During the night these Raccoons appear to wander about, in quest of food, perhaps, to an extent that is almost surprising, so that their tracks can be seen in great numbers in various places, as, even in the dry season, the peculiar tenacity of the soil retains the impression made by their feet, almost as if it were the moulding-sand of the founder.

They are, however, very often observed near the water-courses, are fond of frogs, fish, &c., and their tracks are most likely to be seen in the neighbourhood of streams, even when they are partially dried up, and present only a water-hole here and there.

We have no further knowledge of the habits of this species than the information given in the works of BUFFON, SCHOMBURG, D'AZARA, RENGGER, WAGNER, and the Prince of NEUWIED. In Guiana it is found on the sea-coast; in Brazil and Paraguay, in the bushes and forests, near the rivers and lakes. Besides crabs, it eats birds, eggs, fruits, and is especially fond of sugar-cane. In two individuals that had been tamed, RENGGER did not observe the peculiarity that they dipped their food in the water. SCHOMBURG (Ann. Nat. Hist., iv. 434), however, mentions this habit of others which he saw.

In giving this account of the Crab-Eating Raccoon, we are not entirely without some doubts as to whether the animal found in Brazil and other parts of South America, may not be different from the one in Mexico, Texas, and California. We have, however, inclined to the conclusion that they are the same species, and this the more readily, as the Common Raccoon (*P. lotor*) has a range from Texas to quite a high northern latitude.

GEOGRAPHICAL DISTRIBUTION.

From South America, beyond the tropic, to the shores of the Gulf of Mexico, and on the west as far as California, this species is distributed, but is probably most abundant within the tropics. WAGNER states that it is found from the Caribbean Sea to the 26th parallel of south latitude; BUFFON and SCHOMBURG inform us it exists in Guiana, and we learn from Prince NEUWIED that it inhabits Brazil; while RENGGER and D'AZARA mention its occurrence in Paraguay.

GENERAL REMARKS.

The figure of the Crab-Eating Raccoon, given in our plate, was made by J. W. AUDUBON in the British Museum, from a specimen procured in Mexico or California.

Our description was taken from another specimen in the Charleston

College Museum. This may account for any slight differences between the figure and description.

We have not possessed opportunities of instituting a careful comparison between this animal and *Procyon Lotor ;* they appear, however, to be specifically distinct.

[Thus far we have endeavoured to describe the forms and give the habits of the quadrupeds figured in our work; we will now append some descriptions, and a list of those species we have not been able to portray, but which deserve to be noticed, as belonging to the "Quadrupeds of North America," and necessary to complete the list.]

MEPHITIS ZORILLA.—Gmel.

CALIFORNIAN SKUNK.

(Not figured.)

M. Fronte macula ovali alba insignita; maculâ albâ ad tempus utrumque, strigis quatuor albis, interruptis in dorso et lateribus, caudæ apice albo.

CHARACTERS.

An oval spot of white on the forehead, and a large spot on each temple; four interrupted white stripes on the sides and back; tail broadly tipped with white.

SYNONYMES.

MEPHITIS ZORILLA.—Licht; Darstellung neue, oder wenig bekannter saügethiere,
          1827–1834.   Berlin, tafel xlviii., fig. 2.
LE ZORILLE.—Buffon, Hist. Nat., t. xiii., p. 302, table 41.
MEPHITIS BI-COLOR.—Gray, Loudon's Mag., vol. i., p. 581.
     "   ZORILLA.—Illiger.

DESCRIPTION.

In form, this species may be said to be a small image of the Common Skunk (*M. Chinga*).

Head, short in proportion; ears broad, rounded, clothed with hair on both surfaces; palms naked; nails short, grooved beneath, and slightly hooked; whiskers short and scattering; fur soft, like that of the domestic cat, and composed of two kinds of hair, the under hairs being soft and woolly, the others longer, interspersed among them. On the tail the hair is very coarse, and, toward the extremity, rigid.

COLOUR.

There is a white patch on the forehead, and also between the eye and ear, extending beneath the ear to the middle of the body; another white stripe rises behind the ear, and runs parallel with the foregoing. These stripes are not quite uniform on each side; the body is spotted with white, forming three nearly uniform bars across the back. There are two white spots near the insertion of the tail, on the sides and rump. The white markings are set off by the colour of the remaining portions of the body, being blackish brown, very dark on the head and ears, a little lighter near the flanks.

Tail brownish black, tip (for about three inches) white.

DIMENSIONS.

	Inches.
From point of nose to root of tail, - - - - -	$11\frac{1}{2}$
Tail (vertebræ), - - - - - - -	6
" (to end of hair), - - - - - - -	10
Shoulder to point of longest nail of fore-foot, - - -	5
Height of ear (posteriorly), - - - - - -	$1\frac{4}{10}$

HABITS.

The habits of the present animal are only partially known; it is said to retreat to holes in the earth, or live under roots of trees, in the crevices of rocks, &c. It feeds upon insects, birds, and the smaller quadrupeds.

This Skunk, as we are moreover informed, is able to make itself so offensive that few persons are disposed to approach or capture it, rather keeping aloof, as from the Common Skunk of our Atlantic states, so well known for its "perfume."

GEOGRAPHICAL DISTRIBUTION.

This species was found to be rather abundant, by J. W. AUDUBON and J. G. BELL, in California; it was also found in Texas by the former. DEPPE had discovered it previously in California, in 1820, or thereabouts.

GENERAL REMARKS.

The Zorilla was described by BUFFON (Hist. Nat., tom. xiii., p. 302) as a species existing in South America; his figure, however, bore consider-able resemblance to an African species (*Viverra Striata* of SHAW). Subsequently Baron CUVIER bestowed great attention on this genus, and came to the conclusion that all the American Skunks were mere varieties of each other.

As far as the endless varieties of our Atlantic Species (*M. Chinga*) are concerned, he was correct; but he was greatly in error in regarding the South American, Mexican, and Californian Skunks as being all of one species, for they differ greatly, not only in size, form, and internal organization, but also in colour.

Besides, many species of *Mephitis* present scarcely any variations in colour. The *Mephitis Chinga* seems to be like *Lepus callotis*, the Mexican hare, and *Lynx Rufus*, the bay lynx, a species that may be regarded as an exception rather than a type of the characteristic of the species.

CUVIER came to the conclusion, whilst pursuing his investigations, that BUFFON, in his Zorilla, had described the above named African species; but it now appears that BUFFON was correct, that his specimen came from America, and that the species is found within our limits, on the western coast: therefore we restore his specific name of Zorilla (Le Zorille) as a synonyme.

CANIS (LUPUS) GRISEUS.—Rich.

American Grey Wolf.

(Not figured.)

L. magnitudine canis lupi, cranio lato, gula caudaque villosis, pedibus latis, colore cinereo nigroque notato.

CHARACTERS.

About the size of the black and white wolves; skull broad; neck and tail covered with bushy hairs; feet broad; colour dark brindle grey.

SYNONYMES.

Grey Wolf.—Cook's Third Voyage, vol. ii., p. 293.
    "   "   Lewis and Clarke, vol. i., pp. 206, 283.
Common Grey Wolf.—Schoolcraft's Travels, p. 285.
Canis (Lupus) griseus.—Sabine, Franklin's Voy., p. 654.
    "   Lupus.—Parry, First, Second, and Third Voyages.
    "   "   Harlan, Fauna Americana, p. 81.
    "   "   Godman, American Nat. Hist., vol. i., p. 255, fig. 1.
    "   (Lupus) occidentalis.—(Var. *a*.) Lupus Griseus, Rich. Fauna Borealis Americana, p. 66.
    "   occidentalis, Common American Wolf.—De Kay, Nat. Hist. of N. Y., p. 42, plate 27, fig. 2.
Canis lupus.—Emory, Mass. Report, 1838, p. 26; 1840, p. 28.
Lupus Gigas.—Townsend, Proc. Acad. Nat. Sci., Phila.
Giant Wolf.—Col. G. A. McCall, U. S. A. (letter to Rev. John Bachman, see *infra*).
Lobo or Lovo.—Mexicans and Texans.

DESCRIPTION.

The American Grey Wolf bears a very striking resemblance to the European Wolf. There are, however, some differences which appear to be permanent, and which occur in all the varieties of American Wolves; the body is generally more robust, the legs shorter, and the muzzle thicker and more obtuse in the latter.

We have examined a number of European Wolves (see vol. ii., p. 162, White American Wolf), and although there were great differences between various specimens, we were not able to satisfy ourselves that the American Wolf is the largest, as is supposed by RICHARDSON. We regard them as generally about the same size, and as exhibiting only *varieties*, not specific differences. The body of the American Grey Wolf is long, and rather gaunt; muzzle elongated, and somewhat thicker than that of the Pyrenean Wolf; head thick; nose long; ears erect and conical; eyes oblique—as is the case in all the true wolves—pupil of the eye circular; tail straight, and bushy. The animal does not curl it over the back, like a dog.

Behind the cheek there is a bunch of hairs, which look like a collar. The hairs are of two kinds, the longer coarse and rather rigid, the under fur soft and woolly; whiskers very few, and coarse and rigid; nails long, slightly arched, and, in the specimen from which we describe, considerably worn, as are also the teeth.

COLOUR.

The long hairs, from their roots, for one third of their length, are yellowish white, then a broad band of dark brown follows, succeeded by yellowish brown, and the tips are black. The under fur is ashy brown. On the under surface the long hairs are white nearly to the roots.

The general appearance of the upper surface is dark brindled grey, with an indistinct dorsal line a little darker than the colour of the sides.

The under parts are dull white.

Nostrils black; from the nose towards the eyes, reddish yellow. The outer surface of the ears, and outsides of hind legs, from the hip to the knee joint, are also reddish yellow. The whiskers are black.

DIMENSIONS.

	Feet.	Inches.
Length from point of nose to root of tail,	4	
" of tail (vertebræ),	1	1
" " to end of hair,	1	5
Height of ear,		4
Breadth "		3
From point of nose to end of skull,		11½
" eye to point of nose,		5
" shoulder to longest nail,	2	4
Longest upper canine tooth,		1½

Length of the hair on the back, 3 to 4 inches.

The above description and measurements were taken from a specimen in the possession of our distinguished friend, Prof. KIRTLAND, of Cleveland, Ohio. The animal was killed in February, and was, of course, in full winter pelage.

HABITS.

We have given the general habits of the wolves in our second volume (see pp. 126, 156, 240), and now will favour our readers with the following letter from Col. GEO. A. McCALL, U. S. A., which will be found very interesting. It will be perceived that the Colonel thinks the Giant Wolf, or Lobo, a distinct species. We have, however, thought it best to give Mr. TOWNSEND's name (*L. Gigas*) as a synonyme; and we have also appended to our list of names for the grey wolf, the common one of *Lobo*, or *Lovo*, used by the Mexicans and Texans, although it has been thought by some naturalists that the *Lobo* was a distinct species from any other wolf, and by the Mexican Rancheros it is described as tawny, and with a head like that of a lion.

"The Rev. JOHN BACHMAN, D. D. :

"Dear Sir,—On meeting you in Philadelphia, a few days ago—at which time I passed a most agreeable evening, in company with yourself and other distinguished naturalists, whom your arrival had brought together—you related to us in glowing language a variety of anecdotes, illustrative of the character and habits of different families of our Fauna ; and, in the course of conversation, you inquired whether I had met, in the west, the *Giant Wolf* of N. America. I then mentioned some incidents, which occurred at Fort Gibson several years ago, exemplifying the greater fleetness, power of endurance, and courage of this species, when compared with the common wolf. As you were pleased to pronounce the 'long yarn' worth re-spinning, I, herewith, agreeably to your request, give you my reminiscence of the facts.

"The position of Fort Gibson is, as you are probably aware, near the junction of the Neosho with the Arkansas river ; here, in the angle thus formed, is a prairie of some extent, which used to furnish, for the amusement of those who were fond of the *course* with greyhounds, at any time, a *start* of a prairie wolf (*L. Latrans*), not unfrequently a common wolf (*L. Occidentalis*), and, now and then, the giant wolf himself (*L. Gigas*). The last was easily recognised, even at a distance, by his shyness and his fleetness ; and, as he was generally upon the lookout, it seemed impossible for two or three horsemen, with half a dozen greyhounds, to approach within half a mile of him before he showed a straight tail ; and then his great speed always enabled him to reach cover before the dogs, notwithstanding that two or three of them were of high blood and great fleetness, could overcome the gap which, at the start, separated them from the chase ; and thus the sportsmen, after several killing rides, had always found themselves foiled by the watchfulness and the superior speed and bottom of this wolf.

"After a hard and unsuccessful race of this kind, several officers were one day returning home, when in passing the farm of a Cherokee Indian, they were told by him that a wolf of this description was in the habit of frequenting the grounds about his house, almost nightly ; that he had committed numerous depredations, but that such was his cunning that he had eluded all efforts to kill or capture him. Being assured that a fresh trail might be struck at this point, any morning at daylight, the officers determined to try the fellow's bottom with the fox-hounds. Accordingly, a few nights afterwards—the moon having risen about one o'clock—a party was in the saddle, as soon as they could see upon the prairie, and on their way to the Cherokee's house, which was about seven miles from the Fort. They proceeded leisurely, and reached their destination about three o'clock, purposing to let their horses and dogs rest until daylight, before entering on the chase : the pack, I should mention, consisting of half a dozen fox-hounds, and two or three half-curs, the latter being fleeter and more courageous than the former. It so happened, however, that the dogs, not being coupled, struck the trail close to the house, just as they arrived ; and away they went with a cry, and at a pace which showed that the giant was right before them. For some time the wolf kept within the narrow strip of covert which borders the Bayou Menard, and thus the horsemen were enabled, by a good moonlight, to keep parallel with him on the open plain. But the wolf finding at length that the cover afforded him no security from his pursuers, and trusting to the lightness of his heels, dashed boldly into the prairie, and made a straight course for the hills on the opposite side, at the distance of about three miles. Here he again took cover ; but he was not allowed much time for repose, as the dogs were soon upon him, and the covert which here bordered the Neosho, being like that of the Bayou, narrow, he was soon forced to leave it and the hills, and again take to the plain. In this way the wolf made several bold dashes; running from one cover to another in a straight course of from one to three miles over the plain ; and it was not until half past eight o'clock, A. M., that he was brought to bay. The *denouement* was brought about in this way : the wolf was at last drawing near to cover after one of the open dashes I have mentioned—his speed, to be sure, much abated, and the hounds and horsemen within sight, behind him—when he entered a large field that lay between him and the thicket he wished to gain. The field he soon crossed ; and a good cover, with running water, was within a few yards of him. He knew the grounds well ; but he did not calculate accurately the amount of strength necessary to clear the fence, which here was much higher than on the side where he had entered. Without pause, therefore, he boldly dashed at the obstacle which now alone separated him from all he stood so much in need of ; but as he made the leap his head struck the topmost rail, and he rolled backwards, heavily, upon the ground. Here a shout of triumph from the hunters, who were within view and had witnessed his fall, broke upon his ear ; and now he aroused all his remaining energies for one prodigious effort to effect his escape ; nature, however, was too nearly exhausted to meet the call, and he fell prostrate upon the ground. Horses and hounds were the next moment closing around him, he gained the fence corner, and then turned upon his pursuers.

" A desperate fight ensued—one or two large and powerful half-hound half-cur dogs, in quick succession, rolled away before him, as he dashed against them with his heavy chest and shoulders. Time after time they returned to the charge, for the dogs had their mettle well aroused and were confident of victory, although each moment seemed to diminish the chances in their favour. With each successive round, dog after dog recoiled more or less injured by a quick and violent snap of the giant's jaws—here, on the right, sat a poor, inoffensive looking hound, whose excitement had led him into the depth of a contest for which nature had never intended him, now writhing in agony, and howling most piteously, his long twisted ears drooping lower than ever, while he cast a furtive glance at his lacerated back and shoulders, just released from the jaws of the giant wolf—there, on the left, lay sprawling, another, whose case seemed even more hopeless than the first.

" During the *melée* several pistols had been drawn, to despatch the wolf and save the dogs ; but such was the intricacy of the affair, such the incessant change of position of the combatants, constantly interlocked, that the chances of killing the dogs by a shot were greater than of saving them ; and this continued until dogs and wolf, both, were exhausted, when the latter was knocked on the head with a heavy club. And thus fell the giant wolf, after a run of five hours and a half.

" In this description I have gone much into detail ; but my only desire was to illustrate what I fully believe to be the fact, viz. that the strength, fleetness, and endurance of this wolf are much greater than those of the common wolf, which was never known, in that country, to make anything like such a run as did this fellow. Indeed, it is only necessary to look at the large leg bone, the strong back, the deep shoulder, and broad chest of this wolf, to be satisfied of his superiority to the other, in the qualities I have enumerated. I am also inclined to think that he is more resolute, and not so easily cowed as the other species ; and in support of this opinion, I proceed to the adventure that occurred to Lieut. HOSKINS, with one of this species.

"A few weeks after this, Lieutenant CHAS. HOSKINS, of the 4th Regt. of Infantry, who, being a bold rider and an ardent hunter, was one of the chief actors in the scene I have just described, had a severe encounter with a giant wolf, which I will endeavour to relate as he described it to me.

"He had mounted his horse just before sunset, one day in June, to breathe for an hour, the fresher air of the prairie, and had ridden at a leisurely pace about three quarters of a mile from the fort—his dogs, four or five greyhounds, were following listlessly at his heels, dreaming as little as himself of seeing a wolf—when on a sudden, from a small clump of shumach bushes, immediately at his side, there sprang an enormous giant wolf. By one of those instinctive impulses which it is difficult to describe, horse and dogs were launched upon him before an eye could twinkle. The wolf had but a few yards the start ; and under such circumstances, although the fleetest of his congeners, he stood no chance of escaping from his still fleeter enemies ; in fact, before he had run fifty yards he was caught by the flanks and stopped. Here a most furious fight commenced : it is a well known fact that the greyhound is sometimes a severe fighter, owing to his great activity and his quick, slashing snap, and HOSKINS's dogs were, in addition, in the habit of coursing the prairie-wolf during the fall and winter months, on which occasions the affair was very generally, after a short chase, terminated in about one minute, by the victim having his throat and bowels torn into ribands. This, however, was a different affair ; they had encountered an ugly customer, and the battle was long and of varied aspect. Sometimes the wolf would break entirely clear from the dogs, leaving several of them floored ; again, however, within a few yards he would be checked, and the battle be resumed ; so that during a long struggle there was little change of ground.

"The fight was continued in this way, the prospect of victory or of defeat frequently changing, until both parties were quite exhausted.

"And now, here lay the wolf in the centre, with his tongue hanging from his jaws ; and at the distance of a few feet, the dogs around him, bleeding and panting for breath. At this juncture, HOSKINS, who had not even a penknife in his pocket, was unable to terminate the affair ; he sat upon his horse, a silent and admiring spectator of the strange scene. At length, when he thought his dogs had somewhat recovered their breath, he called on them to return to the charge. Old *Cleon*, a black dog of great strength and courage, was the only one who obeyed the summons—he sprang fiercely at the wolf's throat ; the latter, however, who had risen to his feet, by a well timed snap, seized Cleon by the neck and hind head, and retaining his hold, was grinding away on the poor fellow's skull with his immense jaws. This was too much for any hunter to witness—a favourite dog held helpless, in a grip that threatened very speedily to end his days. HOSKINS was an experienced hunter, and a very cool and determined man—poor fellow, he afterwards fell, fighting most gallantly, at the battle of Monterey, Mexico : on this occasion he sprang from his horse and seized the wolf by the hind leg, and by a violent jerk caused him to release the dog, but only to find, in less than an instant, the jaws of the monster clamped upon his own leg. He told me, the following day, that he plainly felt the jar as the wolf's large canine teeth clashed against each other in the calf of his leg, so powerful was the snap of his jaws.

" The wolf, however, made no effort to shake or lacerate the wound ; at the same time it occurred to the hunter that this would be the only effect of any exertion on his own part to extricate his limb ; and therefore, with the wolf's hind leg in his right hand, and his left leg in the wolf's jaws, he stood perfectly quiet, while poor Cleon, whose head was covered with blood, lay before him, apparently more dead than alive.

" In a moment, however, Cleon recovered and raised his head ; and then his master spoke to him again. Promptly the old fellow obeyed the call, and this time he made good his hold upon the wolf's throat ; whereupon our hunter's leg was at once released. The other dogs now, having pretty well recovered their breath, also re-attacked the wolf ; and this round so disabled him that the affair might be considered as decided. The dogs, however, had all been severely handled, and were again so completely blown that they were unable to make an end of the combat by killing him outright. At this juncture a Cherokee boy, who was on his way across the plain, came up ; but neither had he a knife nor any other weapon. HOSKINS then, as his only resource, unbuckled the reins of his bridle (his horse, well used to such scenes, was quietly feeding, close by), and making of these a slip-noose, he, with the assistance of the boy, got this over the wolf's head, when pulling on the opposite ends, they succeeded in strangling the already exhausted animal. After resting with his dogs a little while, HOSKINS was enabled to mount his horse and return home, with all of them except poor Cleon, who was so much

exhausted as to be unable to keep his legs. A light wagon was immediately sent out for him, and the old dog was received at the fort in
triumph, together with the body of his vanquished adversary. He was,
nevertheless, laid up in hospital for several days, as was his master, whose
leg became inflamed, and prevented his mounting his horse again for a
fortnight.

" The next morning I saw the wolf hanging by the heels, at the front of
the piazza of HOSKINS's quarters ; and he was, beyond all comparison, the
largest wolf that I ever laid eyes upon. His dimensions were taken at
the time ; but I have no memoranda, and I will not venture to speak from
memory.

" The colour and general appearance, however, of these two specimens
(the skins of which were preserved) were, I very well recollect, alike ;
viz. a mixture of rusty black and grey about the head, back, and flanks,
interspersed with a yellowish rusty brown. But the striking marks of
distinction were the large size and the breadth of the head, and the small-
ness of the tail, when compared with other species ; the tail was decidedly
short and scant of hair : the head was very remarkable—I speak of it as I
saw it in the flesh—the front view, taking in what would be included
within a line, drawn between the ears, and two others from those to the
point of the nose, presented very nearly an equilateral triangle ; the head
of the common wolf being much more ovate. Had the skull been stripped
of its integuments, I doubt not it would have shown, to a certain degree,
a corresponding enlargement in the occipital region.

" I feel no hesitation in asserting that these wolves were of the species
recently described by Mr. TOWNSEND as *L. Gigas;* for I did not at the
time, nor have I at any time since, entertained in my own mind a doubt
of this wolf being a distinct species.

" Without instituting any strict inquiry, from personal examination, as
to species or varieties, I have seen a good deal of the wolves of the west
during some years past, and from a difference I have observed in the manners or character of those I have met with in the field, I incline to the
belief that an additional species, between *L. Occidentalis* and *L. Latrans,*
will yet be satisfactorily established.

" G. A. M."

PHILADELPHIA, *July,* 1851.

ARVICOLA DEKAYI.—Aud. and Bach.

Glossy Arvicola.

(Not figured.)

A. Corpore longo ac tenui : naso acuto ; auriculis et pedibus longis ; vellere tereti ac nitente ; supra fusca, subtus cano-fusco.

CHARACTERS.

Body long and slender ; nose sharp ; ears and legs long ; fur smooth and lustrous ; dark brown above, hoary brown beneath.

SYNONYMES.

Arvicola fulva, Glossy Arvicola.—Aud. and Bach., Jour. Acad. Nat. Sciences, Oct. 5, 1841.

Arvicola Oneida, Oneida Meadow Mouse.—De Kay, Nat. Hist. State of New York, 1842, pt. i. p. 88, plate 25, fig. 1.

"   "   Le Conte, Proc. Acad. Nat. Sciences, Phil., Oct. 25, 1853, p. 406.

DESCRIPTION.

This species presents more distinctive markings than any other of the American Arvicolæ ; its body is less cylindrical, and its nose less obtuse than any of our other species ; its ears are prominent, rising two lines above its smooth, compact fur ; its lower incisors are very long, and much exposed, considerably curved ; tail longer than the head, thinly covered with short hairs ; legs long and slender, giving the animal that appearance of lightness and agility observable in the mouse.

COLOUR.

Incisors yellowish white ; the hairs, which are very short, like those on the pine mouse of Le Conte, are at the roots, on the upper surface, plumbeous, broadly tipped with brown, giving it a bright chestnut colour ; the hairs on the legs and toes are a little lighter, on the under surface the colour is cinereous.

DIMENSIONS.

						Inches.	Lines.
Length of head and body, -	-	-	-	-	-	3	9
" tail, -	-	-	-	-	-	1	4
Height of ear (posteriorly),	-	-	-	-			2½
Length of tarsus, -	-	-	-	-	-		7

HABITS.

We have obtained no information in regard to the habits of this species. DE KAY, who obtained a specimen in the neighbourhood of Oneida Lake, in the state of New York, says that it prefers moist places.

GEOGRAPHICAL DISTRIBUTION.

This Arvicola, according to DE KAY, exists in the western part of the state of New York. Our specimen was received from Mr. FOTHERGILL, who procured most of his specimens, we believe, from St. Lawrence county, New York.

We, however, understood that this individual came from Illinois.

GENERAL REMARKS.

It will be perceived, from the dates of our several publications, that we described this species a year previous to DE KAY; the name we gave it, however (*Arvicola fulva*), is pre-occupied by *Lemmus fulvus*, Geoff, which is an arvicola found in France.

As DE KAY described the same animal, without a knowledge of our previous publication of it, we have named it after that naturalist, and have given his name (*A. Oneida*) as a synonyme.

ARVICOLA APELLA.—Le Conte.

Woodhouse's Arvicola.

(Not figured.)

A. Auribus brevissimis sub pilis occultis, intus et extus pilosis. Pedibus gracilibus, brevibus. Cauda brevi, supra obscurè badia, subtus cinereo-plumbea.

CHARACTERS.

Ears very short, concealed beneath the fur, clothed with hair on both surfaces ; feet slender and short ; tail short, brown above, greyish beneath.

SYNONYME.

Arvicola Apella.—Le Conte, Proc. Acad. Nat. Sci., Phil., Oct. 25, 1853, p. 405.

DESCRIPTION.

Head short and blunt ; ears rounded, very short, slightly hairy, both within and without, entirely concealed under the fur, antitragus short, semi-circular. Legs very short ; feet covered with short, shining hairs ; thumb tubercle furnished with a short, blunt nail ; tail very short.

COLOUR.

Hair dark lead-colour, above tipped with brown, redder on the sides ; beneath grey, inclining to brownish on the chin and throat ; feet pale brownish ; tail brown above, greyish beneath.

DIMENSIONS.

	Inches.	Lines.
Length (including the tail),	4	7
" of head,	1	
" " ears,		2
" " fore leg,		5
" " hind "	1	1
" " tail,		7

HABITS.

This animal was procured in Pennsylvania by Dr. WOODHOUSE, in the cultivated portions of that state, and probably has the same propensities and instincts as the other Arvicolæ of North America.

As Major LE CONTE gave it no common name, we have taken the liberty of calling it WOODHOUSE's Arvicola, after the gentleman who procured it.

ARVICOLA AUSTERUS.—Le Conte.

BAIRD'S ARVICOLA.

(Not figured.)

A. Supra fusco et nigro permixtus, subtus obscurè *schistosus*. Auribus extra pilos extantibus, extus pilosis. Cauda gracili, densè pilosa.

CHARACTERS.

Colour, above mixed brown and black, beneath dark slate-colour, mixed with brown; ears longer than the fur, hairy on the outside; tail slender, thickly clothed with hair.

SYNONYME.

ARVICOLA AUSTERUS.—Le Conte, Proc. Acad. Nat. Sci., Phila. Oct. 25, 1853, p. 405.

DESCRIPTION.

Head large and blunt; ears rounded, longer than the fur, outwardly hairy, inwardly only so on the upper margin; antitragus large, semi-circular; whiskers shorter than the head; feet covered with shining hair; thumb tubercle with a compressed, sharp, hooked nail.

Tail slender, covered with short hairs.

COLOUR.

Hair, above dark plumbeous, tipped with brown and black, beneath dark slate-coloured, mixed with brown, particularly on the breast, the upper and under surfaces of the body being nearly alike; whiskers black and grey; feet grey; tail mixed brown and black above, brownish grey beneath.

DIMENSIONS.

			Inches.	Lines.
Length,			5	5
"	of head,		1	3
"	" ears,			8
"	" fore leg,		1	1
"	" hind "		1	5
"	" tail,		1	4

HABITS.

Of the habits and manners of this species we have no account.
Like the foregoing, it has had no common name bestowed on it by
Major LE CONTE. We therefore have called it BAIRD's Arvicola, as
it was found or obtained by Prof. BAIRD. It inhabits Wisconsin.

ARVICOLA CALIFORNICA.—Peale.

CALIFORNIAN ARVICOLA.

A. Subvariegatus rufescenti-fusco et nigro. Corpore brevi et robusto, pilis speciem hirsutici habentibus revera tamen mollibus et levibus. Auribus sub-magnis, pene sub pilis occultis. Cauda supra fusca, subtus fusco-cinerea.

CHARACTERS.

Body short and thick ; hair long and shining, at the roots plumbeous black, above and on the sides tipped with reddish brown and black ; ears rather large, nearly concealed by the fur ; tail brown above, brownish grey beneath.

SYNONYMES.

Arvicola Californica.—Peale, Zool. Explo. Exped., Mammalia, 46.
"   Californicus.—Le Conte, Proc. Acad. Nat. Sciences, Phila., Oct., 1853, p. 408.

DESCRIPTION.

Body short and thick ; hair rather long, and shining : head blunt ; ears large but almost concealed in the fur, hairy on both surfaces ; feet clothed with short, glossy hair ; tubercle of the thumb furnished with a compressed, blunt nail. Tail round ; whiskers numerous, but slender.

COLOUR.

Hair of the body plumbeous black at the roots, above and on the sides tipped with reddish brown and black, in such a manner as to give it a hirsute appearance ; feet greyish brown ; whiskers black and white.

DIMENSIONS.

	Inches.	Lines.
Length, - - - - - - - - -	5	7
" of head, - - - - - - -	1	3
" " fore leg, - - - - - -	2	
" " hind " - - - - - -	1	5
" " tail, - - - - - - -	2	

ARVICOLA OCCIDENTALIS.—Peale.

WESTERN ARVICOLA.

A. Pilis mollissimis et tenuissimis, extremitatibus superioribus rufis sine ulla nigri admistione, auribus sub-pilis occultis. Cauda, sub-compressa, supra et subtus concolore rufa.

CHARACTERS.

Hair very soft and fine ; ears concealed under the fur, hairy only on the outside. Tail slightly compressed, reddish coloured above and beneath.

SYNONYMES.

ARVICOLA OCCIDENTALIS—Peale, Zool. Expl. Exped., i. c. 45.
     "   "   Le Conte, Proc. Acad. Nat. Sciences, Oct. 25, 1853, p. 408.

DESCRIPTION.

Ears round, entirely concealed under the fur, hairy only on the outside, antitragus rather short ; head blunt ; feet covered with short, lustrous hair ; thumb tubercle with a compressed, sharp nail. Tail slightly compressed.

COLOUR.

Hair dark plumbeous, above tipped with bright rufous without any admixture of black : beneath grey, hair on the feet rufous. Tail rufous, both above and below. Incisors pale yellow.

DIMENSIONS.

	Inches.
Length of head and body,	$4\frac{3}{10}$
" tail,	$2\frac{1}{10}$
" hair beyond tail vertebræ,	$\frac{2}{10}$
" hind foot,	$\frac{5}{10}$
" fore " (from wrist to end of toes),	$\frac{5}{10}$
" head,	$1\frac{2}{10}$

Obtained at Puget's Sound, Oregon, by the United States Exploring Expedition.

ARVICOLA (HESPEROMYS) CAMPESTRIS.—Le Conte.

New Jersey Field Mouse.

A. Supra fuscus, subtus cinereo-fuscus. Capite magno, auribus magnis, ovalibus, obtusis, pilis brevibus sparse vestitis.

CHARACTERS.

Above brown, beneath greyish. Head large ; ears large oval, and thinly covered with hair.

SYNONYME.

Hesperomys Campestris.—Le Conte, Proc. Acad. Nat. Sci., Phila., Oct., 1853. p. 413.

DESCRIPTION.

Hair plumbeous black, above tipped with brown, beneath with cinereous brown, darker about the mouth. Head large ; ears large, oval, blunt, thinly covered, both within and without, with very short, closely adpressed hair. Legs and feet brown. Tail well clothed with tolerably long hair.

DIMENSIONS.

		Inches.	Lines.
Length, -	-	3	4
" of head,	-	1	2
" tail,	-	2	7

HABITS.

"This species was found in the collection of the Academy of Natural Sciences, Philadelphia, and labelled *Mus Campestris*, from New Jersey. The specimens were preserved in alcohol, and therefore scarcely fit to be described ; there was, however, enough to show that they were different from any hitherto described animal." (Le Conte.)

ARVICOLA (HESPEROMYS) SONORIENSIS.—Le Conte.

Sonora Field Mouse.

A. Supra saturate cinereus fuscescente-cano leviter intermixtus, subtus albescens. Capite elongato, auribus magnis. Cauda modica.

CHARACTERS.

Above, dark grey slightly mixed with brownish ; breast whitish. Head long and pointed ; legs large ; tail moderate.

SYNONYME.

Hesperomys Sonoriensis.—Le Conte, Proc. Acad. Nat. Sci., Phila., Oct., 1853, p. 413.

DESCRIPTION.

Hair above dark cinereous or slate-colour, slightly mixed with brownish grey, more thickly on the head, nose, and behind the ears, and with grey on the sides ; beneath whitish, except on the throat, which is mixed slate-colour and whitish. Head elongated, pointed ; ears large, oval, hairy both within and without, and with a distinct, narrow grey margin. Feet covered with short, whitish brown hair. Tail moderate, above dark brown, beneath paler.

DIMENSIONS.

		Inches.	Lines.
Length, - - - - - - - - -		3	3
" of head, - · - - - - -		1	2
" ears, - - - - - - -			4
" fore leg, - - - - - -		1	
" hind " - - - - - -		1	8
" tail, - - - - - - -		1	9

" Resembles in some degree the *H. Leucopus.* Collected by the Boundary Commission, under Major Graham." (Le Conte.)

ARVICOLA RUBRICATUS.—Rich.

Red-Sided Meadow Mouse.

A. Supra obscurè plumbeus ; subtus pallidè cinereus, lateribus miniatis, caudâ breviusculâ, pollice minimo.

CHARACTERS.

Back slate-coloured, belly ash-coloured, sides nearly scarlet, tail rather short. Thumb of fore foot rudimentary. Size a little greater than that of the common domestic mouse.

SYNONYME.

Arvicola rubricatus.—Rich. Zool. Beechey's Voy., Mammalia, p. 7.

The above are the characters of a meadow mouse, which burrows in the turfy soil on the shores of Behring's Straits, drawn up from Mr. Collie's notes. In the colours of its fur, and dimensions, it most resembles the *Arvicola œconomus* (Pall. glir. n. 125., pl. 14, A.), and appears to be quite distinct from any American meadow mouse hitherto described. There is no specimen in the collection. (Richardson.)

PEROGNATHUS PENICILLATUS.—Woodhouse.

Tuft-tailed Pouched Rat.

CHARACTERS.

Above yellowish brown, beneath white; tail longer than the head and body, penicillate, with bright brown hair.

SYNONYME.

Perognathus Penicillatus.—Woodhouse, Proc. Acad. Nat. Sciences, Phil., Dec., 1852, p. 200.

DESCRIPTION.

Head of moderate size, not easily distinguished from the neck; incisors small and partially exposed, upper ones sulcate in the middle. Nose small and rather pointed, extending some distance beyond the incisors; whiskers light brown, irregularly mixed with black; eyes dark brown, and of moderate size; ears nearly round and moderate, almost naked anteriorly, and covered posteriorly with fine brown fur; the tragus and anti-tragus are quite prominent. The external meatus is protected by a tuft of short, black bristles extending across the ear. Tail about one inch and a quarter longer than the head and body, round, gradually tapering, and covered with hair; on the superior and middle portion commences a row of long, silky hairs, which gradually increase in width until they form a tuft at the end. Fore legs short, feet small, with four well developed toes and a short thumb, which is armed with a nail; palms naked. Hind legs and feet long, having five toes, terminated by nails. Feet and toes covered with fine short fur; soles naked. The fur longer on the back than on the belly; it is thick, soft, and silky.

COLOUR.

Incisors yellow, top of head and back dark yellowish brown, lighter on the sides; fur at base light ash colour. Throat, belly, vent, fore legs, and inner portions of thighs white. The white commences at the nostrils, and forms a well marked line to the thighs, and extending down to the heel, leaving the front of thigh white, the remainder and outer portion light yellowish brown; feet white. Under portion of tail white, above dark brown; the long hair of the tail is a rich brown.

DIMENSIONS.

	Inches.
Length from tip of nose to root of tail, - - - -	3·5
" of tail (vertebræ), - - - - - -	3·7
" " ear anterior, - - - - - - -	·3
" " whiskers, - - - - - - -	1·7
" " os calcis, middle toe nail, - - - -	1·
Distance from anterior angle of orbit to tip of nose, -	·6¼

GEOGRAPHICAL DISTRIBUTION.

New Mexico, west of Rio Grande.

GENERAL REMARKS.

Of the habits of this animal I know but little. The specimen in my possession is a male, and was procured in the San Francisco Mountain, New Mexico. (WOODHOUSE.)

PSEUDOSTOMA (GEOMYS) FULVUS—Woodhouse.

Reddish Pouched Rat.

CHARACTERS.

Light reddish brown above, beneath whitish. Ears small, round, and covered with thick, short, black fur. Tail long in proportion when compared with others of this genus.

SYNONYME.

Geomys fulvus.—Woodhouse, Proc. Acad. Nat. Sci., Phila., 1852, p. 201.

DESCRIPTION.

Head large, nose broad, covered with short, thick fur, with the exception of a small space at tip and the margins of the nostrils, which are naked. The nose extends a short distance beyond the plane of the incisors. The incisors are exserted, with three convex smooth sides, the exterior broadest, and of a yellowish colour; their cutting edges are even. The upper incisors extend downwards and inwards; the under ones are one-third longer than the upper, and slightly narrower. Ears small and round, covered with short, thick, black fur externally. Eyes larger than is common in this genus. Tail round, thick at base, and gradually tapering. The fore claws are long, compressed, slightly curved, and pointed. The claw on the middle toe is the longest, the fifth is the shortest, and that of the thumb resembles much the claw of the fourth toe of the hind foot, both as regards size and shape. The toes on the hind feet are a little longer and more slender than those of the fore feet; the nails short, somewhat conical and excavated underneath.

COLOUR.

Head, cheeks, back, and sides bright reddish brown, being darker on the top of the head and back. The breast, vent, feet, inner portion of legs and thighs white, slightly inclining to ash; abdomen very light reddish brown; fur at base dark ash colour above, beneath light ash. Edges of cheek pouches encircled with rufous; the long hair of the back extends about one-third the length of the tail. The tail is covered with short, white, silky hairs, terminating in a small tuft. The fore feet above are covered with short, white hair ; the toes on their inner side have a row of long white hairs ; palms naked. Claws are opaque, white for half their extent, the other half transparent ; there is a small, oblong, reddish brown spot in the centre of each. The hind feet are covered above with white hairs, soles naked. The lips, on their inner side, are covered with short, fine white hair, with a band of short, fine, black fur encircling the mouth. Whiskers silvery white.

DIMENSIONS.

		Inches.
Length from tip of nose to root of tail, - - - -		5·
" of tail (vertebræ), - - - - - -		1·3
" from anterior angle of eye to tip of nose, - -		·7
" " tip of nose to auditory opening, - - -		1·1½
" of os calcis, including middle toe and claw, - -		1·1
" from elbow to end of middle hind claw, - -		1·8
" of middle fore claw, - - - - - -		·4
" " hind claw, - - . - - - -		·2½
" " fur on back, - - - - - -		·2½
" " whiskers, about - - - - - -		1·

GEOGRAPHICAL DISTRIBUTION.

New Mexico, west of Rio Grande.

GENERAL REMARKS.

The specimen in my collection was procured near the San Francisco Mountain, New Mexico, where they were quite abundant.

These Pouched Rats of the genus Perognathus and Geomys I procured whilst attached as Surgeon and Naturalist to the party under command of Capt. SITGREAVES, U. S. Army, exploring the Zuñi, and Little and Great Colorado Rivers of the west. (WOODHOUSE.)

ARVICOLA MONTANA.—Peale.

PEALE'S MEADOW MOUSE.

A. Formâ rotundatâ ; capite magno ; auribus mediocribus et vellere pœne vestitis ; dentibus flavis ; oculis parvis, nigris ; pilis subtilibus sericisque, in dorso brunneis nigrisque intermixtis ; infrà plumbeis. Cauda pedibusque brevi nitente pilo indutis. Mystacibus albis nigrisque : mammis octo, quatuor in abdomine, in pectore totidem.

CHARACTERS.

Form rounded ; the head large, ears moderate and nearly covered with fur ; teeth yellow ; eyes small, black ; hair fine and silky ; that of the back brown and black, intermixed ; beneath lead-coloured ; tail and feet covered with short, glossy hairs ; whiskers white and black ; teats eight in number, four pectoral, and four abdominal.

SYNONYME.

ARVICOLA MONTANA.—Peale, Mammalia and Ornithology United States Exploring Expedition, vol. viii., p. 44.

DIMENSIONS.

Total length 6½ inches, including the tail, which is 1½ inches long.

GENERAL REMARKS.

Our specimen was obtained on the 4th of October, near the head waters of the Sacramento River, in California. (PEALE.)

In relation to *Arvicola Riparia* of ORD, we have concluded that it is identically the same as *A. Pennsylvanica* of that naturalist. We have given an account of this animal at p. 341, Vol. I. We merely mention that it is so much better known as *A. Pennsylvanica* than as *Riparia*, that we would, setting aside the dates of description by Mr. ORD, prefer to let the name of *Pennsylvanica* remain, and for the future consider *riparius* as a synonyme only.

We may further remark, that had we had an opportunity of examining a specimen of the so called "*Arvicola riparius*," from the locality in which Mr. ORD procured his original, before our article on *A. Pennsylvanica* was published, we should have given either the one or the other name as a synonyme. We have lately had a fine specimen of this Arvicola from the locality from whence Mr. ORD obtained his original specimen.

PSEUDOSTOMA CASTANOPS.—Baird.

Chestnut-cheeked Pouched Rat.

(In Stansbury's Report of the Expedition to the Great Salt Lake, p. 313.)

DESCRIPTION.

General colour pale yellowish brown. There is an ample patch of light chestnut on the side of the head and face, deepest above. The dorsal line is not darker than the rest of the fur. Size intermediate between *P. borealis* and *P. bursarius*.

COLOUR.

The colour of the fur above is slightly grizzled, and much lighter than in *P. bursarius ;* beneath paler ; throat, space between the fore legs and arms pale rusty. The chestnut marking on the side of the head is very strongly defined, occupying on each side a nearly circular space of about one and three quarter inches in diameter, with the ear as the centre. These chestnut spaces do not quite meet on the crown and occiput, but leave a rectilinear interval, coloured like the rest of the back, of about one-eighth of an inch in width. On the muzzle, however, from above the eyes the colour of the opposite sides is confluent. The hind feet and toes are thinly covered with whitish hairs, which on the fore feet appear more ferruginous.

The claws are white, but sufficiently transparent to allow the coagulated blood to show through them.

DIMENSIONS.

		Inches
Length to base of tail (approximate),	- - - -	8
" of tail, - - - - - - - - -		$2\frac{5}{8}$
" " hand (along the palm), - - - - -		$1\frac{1}{4}$
" " middle anterior claw, - - - - -		$\frac{1}{2}$
" " hind feet (along sole) from heel, - - -		$1\frac{3}{8}$

HABITS.

This beautiful species was collected by Lieutenant ABERT, on the prairie road to BENT's fork.

The above description and remark we have taken from Prof. BAIRD, with scarcely any alteration.

We have added an English name to the animal.

PSEUDOSTOMA (GEOMYS) HISPIDUM.—Le Conte.

P. Pilis concoloribus rufo-fuscis minus subtilibus tectus, cauda brevi-nuda, auribus obsoletis.

SYNONYME.

Geomys Hispidum.—Dr. Le Conte, Proc. Acad. Nat. Sci., Phila., 1852, p. 158.

DESCRIPTION.

One specimen, Mexico, Mr. Pease's collection. This species differs from all the others in having the fur very coarse and harsh, and entirely of a reddish brown colour. Beneath it is slightly greyish, but the difference in colour is by no means obvious. The ears are not at all prominent, being merely openings in the skin. The whiskers are as long as the head. The upper incisors are broken off, but enough remains to show that they were deeply grooved near the middle of the anterior surface ; it is impossible to determine if there is a second submarginal groove. The tail is completely naked except at the root. The feet are precisely as in the other species of this division of the genus. (Dr. Le Conte.)

DIMENSIONS.

	Inches.
Length from nose to root of tail, - - - - -	11·5
Tail, - - - - - - - - - -	3
Anterior foot to end of claw of third toe, - - - -	1·7
Posterior foot to end of claw of third toe, - - -	1·9

PSEUDOSTOMA UMBRINUS.—Rich.

G. Super umbrinus, subter griseus, gulâ pedibusque albidis, caudâ grisea vestitâ longitudine capitis.

CHARACTERS.

Umber brown on the dorsal aspect, grey below, with white feet and throat, and a grey hairy tail as long as the head.

SYNONYMES.

Geomys Umbrinus.—Rich, Fauna Boreali Americana, p. 202.
      "   "   Dr. Le Conte, Proc. Acad. Nat. Sci., Phila., 1852, p. 162.

DESCRIPTION.

Head large, nose wide and obtuse, and with the exception of the nostrils, covered with fur similar in colour and quality to that on the crown of the head. The nostrils are small round openings, half a line apart, with a furrowed septum, and having their superior margins naked and vaulted ; a narrow, hairy, upper lip, not exceeding a line in width, separates the nostrils from the upper incisors. The whiskers are white, and are shorter than the head. The incisors are much exserted, and are without grooves on their anterior surfaces, which are slightly convex. and of a deep yellow colour. The lips unite behind the upper incisors, so as to form a naked furrow leading towards the mouth, which is rendered more complete by the stiffness of the hairs on each side of it. The cheek pouches are of a soiled buff colour, and are clothed throughout their exterior surface with very short, soft, whitish hairs, which do not lie so close as entirely to conceal the skin. The middle of the pouch is opposite to the ear, and its anterior margin extends forwards to between the eye and the angle of the mouth ; its tip is rounded.

The body, in shape, resembles that of a mole. It is covered with a smooth coat of fur, of the length and quality of that of a meadow mouse ; but possessing more nearly the lustre and appearance of the fur of a musk rat. For the greater part of its length from the roots upwards, it has a blackish grey colour. On the upper and lateral parts of the head, and over the whole of the back, the tips of the fur are of a nearly pure umber-brown color, deepest on the head, and slightly intermixed with chestnut brown on the flanks. The belly, and fore and hind legs, are pale grey, with, in some parts, a tinge of brown.

The sides of the mouth are dark-brown, with a few white hairs inter-mixed. The chin, throat, feet, and claws, are white. The tail is round and tapering, and is well covered with short greyish white hairs ; the hairs on the sides of the fore-feet are rather stiff, and curve a little over the naked palms ; those on the hind-feet are shorter ; the posterior extremi-ties are situated far forward.

DIMENSIONS.

	Inches.	Lines.
Length of head and body, - - - - -	7	
" of head, - - - - - - -	1	8
" " tail, - - - - - - -	1	9
Distance from the end of the nose to the anterior angle of the orbit, - - - - -		9

HABITS.

" Although this animal is not an inhabitant of the fur countries, the above description has been inserted with the view of rendering the account of the genus more complete." (RICHARDSON.)

RICHARDSON received no information respecting its manners or food. The specimen came from the south-western part of Louisiana.

PSEUDOSTOMA (GEOMYS) MEXICANUS.—Le Conte.

P. Mexicanus, mollipilosus, saturate cinereus, supra nigro-tinctus, naso brunneo, cauda mediocri, pilosa, versus apicem subnuda, auribus brevibus, primoribus superioribus medio profunde sulcatis.

SYNONYMES.

Ascomys Mexicanus.—Lichtenstein, Abhandl. Berl. Akad. 1827, 113.
    "   "   Brantz, Muiz. 27.
    "   "   Wagner, Schreb. Saügth. Suppl. 3, 384.
    "   "   Schinz, Syn. Mam. 2, 133.
Saccophorus Mexicanus.—Fischer, Richardson, Rep. Brit. Ass. 6, 156.
    "   "   Syn. Mam. 305.
    "   "   Eydoux, Voy. Favorite, 23, tab. 8.

DESCRIPTION.

One specimen, Mexico, Mr. J. Speakman. Fur very fine, shining, very dark cinereous, above tipped with black, beneath entirely cinereous ; nose and whiskers brownish ; breast and fore-legs slightly tinted with brown. Ears short. Upper incisors with a very deep groove on the middle of the anterior surface. Feet thinly clothed with brownish hair. Tail covered with hair, which is very dense and long at the base, gradually becoming shorter and more scanty, leaving the tip almost naked. (Dr. Le Conte.)

DIMENSIONS.

	Inches.
Length from nose to root of tail, - - - - -	11
" tail, - - - - - - - -	5
Fore-foot to end of middle claw, - - - - -	1·7
Hind-foot to end of middle claw, - - - - -	1·7

SOREX FORSTERI.—Richardson.

Forster's Shrew Mouse.

(Not figured.)

S. Caudâ tetragonâ longitudine corporis, auriculis brevibus vestitis, dorso xerampelino, ventre murino.

CHARACTERS.

Tail as long as the body and square ; ears short and furry ; back brown, belly pale yellowish brown.

SYNONYMES.

Shrew, No. 20.—Forster, Phil. Trans., vol. lxii., p. 381.
Sorex Forsteri.—Richardson, Zool. Jour., No. 12, April, 1828.
"   "   Bachman, Jour. Acad. Nat. Sci., Philadelphia, vol. vii., part ii.,
p. 386.

DESCRIPTION.

Nose, long, somewhat divided at the tip ; ears, hairy, not much shorter than the fur, but still concealed ; body slender ; feet small ; tail long, four-sided ; hair short, fine, and smooth.

COLOUR.

The fur is for two thirds of its length dark cinereous above, tipped with brown ; beneath it is cinereous.

Feet flesh coloured ; nails white.

DIMENSIONS.

							Inches.
Length of head and body,	-	-	-	-	-	-	$2\frac{3}{8}$
" of head,	-	-	-	-	-	-	$\frac{3}{4}$
Height of ear, -	-	-	-	-	-	-	$\frac{1}{8}$
Length of tail, -	-	-	-	-	-	-	$1\frac{1}{2}$
From point of nose to eye,	-	-	-	-	-	-	$\frac{3}{8}$

SOREX COOPERI.—Bach.

CHARACTERS.

Body very small ; nose long ; no external ears ; tail as long as the body ; colour, dark brown.

SYNONYME.

Sorex Cooperi.—Bachman, Jour. Acad. Nat. Sci., Philadelphia, vol. vii., part ii., 1837, p. 388.

DESCRIPTION.

Body very slender, head rather long, and nose thin and pointed ; legs slender and long, especially the hind legs. They are covered with fine adpressed hairs to the extremities of the nails. Tail large and thick for the size of the animal ; flattened on the sides and beneath, rounded above, clothed with fine hair and tipped with a pencil of hairs. The eye is small, but is visible through the fur, and apparently not covered by an integument.

The point of the nose is slightly divided ; there is no external ear, and the transverse auditory opening is completely concealed by the fur.

COLOUR.

Hair cinereous for two thirds of its length above, and tipped with shining chestnut brown ; beneath tipped with ash color ; feet grey ; tail brown above, silver grey beneath.

DIMENSIONS.

	Inches.
From point of nose to tail,	1⅞
Length of tail,	1⅞
From eye to point of nose,	⅜
Length of head,	¼
From heel to middle claw,	₁₆⁷

SOREX FIMBRIPES.—Bach.

CHARACTERS.

No external ears ; tail a little shorter than the body ; feet broad, fringed at the edges ; body dark brown.

SYNONYME.

Sorex Fimbripes.—Bach, Jour. Acad. Nat. Sciences, Philadelphia, vol. vii., part ii., p. 391.

DESCRIPTION.

Nose long and movable, with the tip slightly lobed ; head large and flat. The eye is a mere speck, covered by the common integument, and is found with great difficulty. Whiskers, long, extending considerably beyond the head ; no external ears, and the transverse auditory opening very small ; fore-feet broad, and clothed with short fine hairs extending to the extremities of the nails, the edges on the lower surface considerably fringed beneath the palms with long brownish hairs. Tail of moderate size, square, and gradually tapering to the point.

The fur is considerably longer than in any other of our species of shrew of the same size.

COLOUR.

Teeth yellowish ; whiskers white ; there is a lightish edge around the upper lip ; feet dingy yellow.

The fur on the upper surface is for two thirds of its length, bluish ash, and is tipped with brown, which gives it a changeable brown appearance. Throat and beneath dark fawn colour. Under side of tail buff ; point of tail nearly black.

DIMENSIONS.

	Inches
From point of nose to root of tail, - - - - -	$2\frac{1}{8}$
Length of tail, - - - - - - - -	$1\frac{3}{4}$
From orifice of ear to point of nose, - - - -	$\frac{3}{4}$
" eye to point of nose, - - - - -	$\frac{3}{8}$
" heel to end of middle claw, - - - - -	$\frac{1}{2}$
Breadth of fore-feet, - - - - - - -	$\frac{3}{16}$
Length of whiskers, - - - - - - -	1

SOREX PERSONATUS.—St. Hillaire.

SYNONYMES.

Sorex Personatus.—St. Hillaire, Guerin's Mag. de Zoologie pour 1833, pl. 14.
     "   "   Bachman, Monogr. N. American species of Sorex, Jour.
             Acad. Nat. Sci., Phila., vol. vii., part ii., p. 398.

DESCRIPTION.

Hair reddish brown above, light ash coloured beneath, end of the nose blackish brown above, ears small and concealed in the fur ; tail rather square, one third of the total length of the animal.

DIMENSIONS.

	Inches.
Length to root of tail, - - - - - - -	2
" of tail, - - - - - - - - -	1

HABITS.

We have never seen this shrew. The specimen from which the description was taken by St. Hillaire (translated above) was sent from America by Milbert (1827).

GEORYCHUS GRŒNLANDICUS.—Rich.

Greenland Lemming.

A. Exauriculatus, rostro acuto, palmis tetradactylis hirsutis ; unguibus apice cylindrico producto, lineâ dorsali nigrâ.

CHARACTERS.

Earless ; with a sharp nose ; fore-feet hairy beneath, with four toes, armed with claws, having sharp cylindrical points : a dark stripe along the middle of the back.

SYNONYMES.

Arvicola (Georychus) Grœnlandicus, Greenland Lemming.—Rich, Fauna Boreali Americana, p. 134.

Mouse, Sp. 15.—Foster, Phila. Trans. lxii., p. 379 ?

Hare tailed Rat ?—Pennant, Arct. Zool., vol. i., p. 133 ?

Mus Grœnlandicus.—Richardson, Parry's Second Voy., App. p. 304.

Owinyak—Esquimaux.

DESCRIPTION.

Size—rather less than a rat : head rounded, narrower than the body, tapering slightly from the auditory opening to the eyes ; nose acute. There are no external ears, but the site of the auditory opening is denoted by an obscure transverse brownish streak in the fur. The eyes are near each other and small. The fur on the cheeks is a little puffed up. The upper lip is deeply divided ; lower incisors twice the length of the upper ones ; whiskers long ; body thickly covered with long and soft fur. Tail very short ; the *fore extremities* project very little beyond the fur ; the palms incline slightly inwards, are small, and the toes very short ; both are covered thickly above and below, with strong hairs curving downwards, and extending beyond the claws. The only naked parts on the foot are a minute, flat, unarmed callus, in place of a thumb, and a rounded smooth callus at the extremity of each toe. These callosities do not project forward under the claws, and have no resemblance to the large, compressed, horny, under portions of the claws of the Hudson's Bay Lemming.

The claws are long, strong, curved moderately downwards, and inclining inwards. Soles of the hind-feet hairy, and the hairs project beyond the claws. The hind-feet have five toes, of which the three middle ones are nearly of a length. The hind-claws are slightly arched, narrow, but not sharp at the points; they are thin, hollowed out underneath, and calculated to throw back the earth which has been loosened by the fore-claws.

COLOUR.

The general colour of the upper parts of the body and of the head is dark greyish brown, arising from an intimate mixture of hairs tipped with yellowish-grey and black; the black tips are the longest. and, predominating down the centre of the back, produce a distinct stripe. The ventral aspect of the throat, neck, and body, exclusive of some rusty markings before the shoulders, is of an unmixed yellowish-grey colour, which unites with the darker colour of the back by an even line running on a level with the tail and inferior part of the cheek. The fur both on the back and underneath presents, when blown aside, a deep blackish-grey colour from the tips to the roots. The tail is of the same colour as the body at the root, but the part which projects beyond the fur of the rump is only a pencil of stiff white hairs.

The above is copied, with some alterations, from RICHARDSON's description, which was drawn up from a male, killed August 22, in Repulse bay.

DIMENSIONS.

		Inches.	Lines.
Length of head and body, - - - - -		6	3
" " tail, - - - - - - -			9
" " fore-leg from palm to the axilla, - -		1	1
" " longest fore-claw, - - - - -			4
" " palm of middle-claw, - - - -			6
" " whiskers, - - - - - -		1	4

HABITS.

We refer our readers to the Fauna Boreali Americana for some interesting general remarks on the Lemmings, comparing those of the American continent with European species. We know nothing of the habits of this one.

DIPODOMYS ORDII.—Woodhouse.

Ord's Pouched Mouse.

CHARACTERS.

Light reddish brown above, beneath white ; tail short, and penicillate at the end.

SYNONYME.

Dipodomys Ordii.—Woodhouse, Proc. Acad. Nat. Sci., Phila., 1853, p. 235.

DESCRIPTION.

A little smaller than *D. Philipsii*, Gray ; head and tail shorter, nose long and pointed, extending some distance beyond the incisors ; ears some-what round, the anterior portion almost naked, posteriorly covered with short fine hair.

COLOUR.

Dark reddish brown above ; sides light reddish brown ; fur ash colour at base ; side of the nose, half of the cheek, spot behind the ear, band across the thigh and beneath, pure white ; a black spot at the base of the long whiskers ; a superciliary ridge of white on either side ; the penicillated portion of the tail is formed of long white hairs, with bright brown tips.

DIMENSIONS.

		Inches.
Total length from tip of nose to root of tail, - - -		5
" " of vertebræ of tail, - - - - -		$4\frac{3}{10}$
" " of tail, including hair at tip, - - -		$5\frac{5}{8}$
" " of os calcis, including middle toe and nail, -		$1\frac{5}{8}$
" " of ear, - - - - - - - -		$\cdot45$

GEOGRAPHICAL DISTRIBUTION.

Western Texas.

GENERAL REMARKS.

This animal I procured at El Paso on the Rio Grande, on my way to Santa Fé, whilst attached to the party under the command of Captain L. SITGREAVES, United States Army. I have named it in honour of Mr. ORD, President of this Society. (WOODHOUSE.)

ARVICOLA (HESPEROMYS) TEXANA.—Woodhouse.

CHARACTERS.

Smaller than Mus leucopus, head shorter and more blunt, ears smaller and more round, brown above, and white, inclining to yellowish, beneath.

SYNONYME.

Hesperomys Texana.—Woodhouse, Proc. Acad. Nat. Sci., Phila., 1853, p. 242.

DESCRIPTION.

Head large, blunt. Eyes prominent, and dark brown. Ears large, erect, roundish, oval, blunt, sparsely covered outwardly with short adpressed brown hairs, inwardly with grey. Thumb of fore-feet a tubercle, furnished with a long blunt nail, two middle toes the longest, subequal. Hind-feet furred, with the exception of the sole. Whiskers long.

COLOUR.

Hair dark cinereous above, tipped with pale brown, and dusky, so as to have rather a mottled appearance ; beneath white inclining to yellowish ; the two colours, that is to say above and beneath, tolerably distinctly separated from each other in a straight line. Tail above brown, beneath white ; nose mixed brown and grey, or pale brown. Whiskers black and grey ; legs white on their inner surface only, feet white, the hairs projecting over the nails.

DIMENSIONS.

	Inches.
Total length from tip of nose to root of tail, - · ·	$2\frac{1}{10}$
" " of tail, - · · · · · ·	$2\frac{1}{10}$
" " of head, - · · · · · ·	$1\frac{1}{10}$
Height of ear, - · · · · · · ·	$\frac{4}{10}$
Breadth of ear, - · · · · · · ·	$\frac{3}{10}$
Fore-legs, - · · · · · · · ·	1
Hind-legs, - · · · · · · · ·	$1\frac{6}{10}$

GEOGRAPHICAL DISTRIBUTION.

Western Texas.

GENERAL REMARKS.

I procured this little animal on the Rio Grande near El Paso, whilst attached to the party under the command of Captain L. SITGREAVES, U. S. Topographical Engineers, on our way to explore the Zuni and Colorado rivers. Of its habits I know nothing. My attention was called to this animal by Major LE CONTE, who has been for some time engaged in the study of the mice of our country. (WOODHOUSE.)

SCALOPS ÆNEUS.—Cassin.

Black-clawed Shrew Mole.

SYNONYME.

Scalops Æneus.—Cassin, Proc. Acad. Nat. Sci., Phila., 1853, p. 299.

DESCRIPTION.

Upper jaw, after the two incisors, having on each side seven false molars, which are pointed and nearly equal, except the last, which is double the size of either of the others, and has a small exterior basal lobe. Molars three; the first with four external lobes, the anterior being very small, the second large and pointed, the third short, blunt, and deeply emarginate, the fourth lobe also blunt and short; besides these the first molar has one interior and one posterior lobe, second molar with three short external lobes, the intermediate one emarginate; also two interior large and pointed, and one posterior similar to the interior lobe; third molar with two short external lobes, the posterior one emarginate, and two interior lobes and one posterior lobe.

Lower jaw with two incisors on each side, the anterior of which is the shorter; these are followed by six false molars, which are pointed and nearly equal in size, except the last, which is much larger and furnished with a minute posterior lobe at the base. Molars three, each deeply sulcate on the external surface and composed of two large external lobes and three smaller and shorter internal lobes.

First and fifth toes of fore-feet equal, second shorter, first and fifth toes of the hind-feet equal, other three nearly so.

COLOUR.

Entirely shining, brassy brown, very glossy, and in some lights appearing to be almost metallic; darker on the top of the head, and lighter and more obscure on the chin and throat; nose dusky; feet brownish; nails and first joint of the toes black; palms dusky; soles of the hind-feet dark brown; tail light brown, thinly furnished with scattering bristles.

DIMENSIONS.

			Inches.
Total length (of specimen in spirits), about -	-	-	-	5
" " of head, -	-	-	-	2
" " of fore-feet, -	-	-	-	1·15
" " of hind-feet, -	-	-	-	1·40
" " of tail, -	-	-	-	1·25

GENERAL REMARKS.

This is the most beautiful species of mole yet discovered in America, and exhibits almost the brilliancy of colour which distinguishes the remarkable South African animals which form the genus Chrysochloris, of this family.

A single specimen, apparently fully adult, is in the collection of the Exploring Expedition, labelled as having been obtained in Oregon. In its dentition and otherwise it is a strict congener of *Scalops Townsendii*, but is much smaller and of a different color. Its black claws are especially remarkable, and distinguish it from all other species of the genus.

(CASSIN.)

SCALOPS LATIMANUS.—Bach.

Texan Shrew Mole.

SYNONYME.

Scalops Latimanus.—Bach, Boston Jour. Nat. History, vol. i., p. 41.

DESCRIPTION.

Larger than the common shrew-mole, intermediate in size between *S. Townsendi* and *S. Breweri*. Hair longer and thinner than in either of the other species, and slightly curled. Palms larger than in any other known species. Tail naked.

COLOUR.

Colour nearly black.

DIMENSIONS.

	Inches.	Lines.
Length to root of the tail, - - - - -	7	7
" of the tail, - - - - - -		10
Breadth of the palm, - - - - - -		10
" of the tarsus, - - - - - -		7

GEOGRAPHICAL DISTRIBUTION.

Mexico and Texas.

MUS LE CONTII.—Bach.

LE CONTE'S MOUSE.

M. Supra rufo-fuscus, subtus albo-flavus ; cauda corpore breviore.

CHARACTERS.

Tail shorter than the body, reddish brown above, light fawn beneath.

SYNONYMES.

MUS LE CONTEI.—Aud. and Bach, Jour. Acad. Nat. Sci., Phila., vol. viii., pt. ii.,
p. 306.
REITHRODON LE CONTEI.—Le Conte, Proc. Acad. Nat. Sci., Phila., Oct. 1853,
p. 413.

DESCRIPTION.

About half the size of a full grown mouse. Its body is covered by
a very thick coat of soft fur and coarser hairs intermixed. The upper
fore-teeth are deeply grooved. The head is of a moderate size ; the forehead so much arched as to present nearly a semicircle. Nose rather sharp,
with a caruncle beneath each nostril pointing downwards. Whiskers
shorter than the head. Ears round, moderate in size, and slightly protruding beyond the long fur, nearly naked ; a few hairs are sprinkled
along the inner margins. The legs are short and rather stout ; feet
covered with short adpressed hairs ; nails long and but slightly hooked ;
adapted to digging. The rudimentary thumb is armed with a blunt nail.
The tail, which is round, is sparsely clothed with hair.

COLOUR.

Teeth yellow ; eyes black ; nails light brown ; whiskers white and
black. The fur on the back and chest is plumbeous at base, tipped with a
mixture of reddish brown, and dusky, giving it a dark reddish-brown
appearance. The lips, chin, and feet are a soiled white. On the throat,
belly, and under surface of the tail, the fur is cinereous ; at the roots tipped
with fawn colour. Upper surface of tail brown.

DIMENSIONS.

	Inches.	Lines.
Length of head and body, - - - - -	2	6
" of tail, - - - - - - -	2	
Height of ear, - - - - - - -		1½
Length of tarsus, - - - - - - -		5

GENERAL REMARKS.

The specimen from which the above description was taken was procured in Georgia by Major LE CONTE.

MUS MICHIGANENSIS.

M. Buccis flavis, corpore supra fusco-canescente, subtus albido.

CHARACTERS.

Cheeks yellow, body light greyish-brown above, whitish beneath.

SYNONYME.

Mus Michiganensis.—Aud. and Bach, Jour. Acad. Nat. Sci., Phila., vol. viii., pt. ii., p. 304.

DESCRIPTION.

The head is of moderate size at base, gradually tapering to a sharp-pointed nose. The eyes, which appear to be rather smaller than those of the white-footed mouse, are placed farther forward. Whiskers the length of the head. The ears on both surfaces are so sparingly clothed with short hairs as, without close examination, to appear naked. Legs short and slender, covered with hair to the extremities of the toes. Soles naked. On each fore-foot there are four toes, with a rudimental thumb, protected by short but rather sharp nails. The hind-feet are pendactylous. The tail, which is round, is clothed with rather short hairs. Mammæ, six pectoral and four abdominal. The fur on the whole body is very short and smooth.

COLOUR.

The incisors, which are small, are yellow. The whiskers are nearly all white ; a few immediately below and above the eyes being black. On the cheeks there is a line of yellowish fawn colour running along the sides to the neck. The feet, nails, ears, and tail are light brown. The hairs on the upper surface are light plumbeous at the roots, and tipped with light brown and black. On the throat, inner surface of the thighs, and on the abdomen, they are yellowish white. There is no distinct line of demarcation between the colours of the back and under surface, nor does the white extend along the sides, as in the white-footed mouse.

DIMENSIONS.

		Inches.	Lines.
Length of head and body,	- - - - -	4	
" of tail, -	- - - - - - -	2	6
" of tarsus,	- - - - - - -		5
Height of ear, -	- - - - - - -		4

GENERAL REMARKS.

This species bears some resemblance in size and colour, both to the common house mouse (*M. musculus*) and the white-footed mouse (*M. leucopus*). The colour on the back resembles the former, and on the under surface the latter. Its tail is considerably shorter than either, and its ears less naked and much smaller than those of *M. leucopus*. Neither has it the white feet so characteristic of that species.

PEROGNATHUS (CRICETODIPUS) PARVUS.—Peale.

P. Capite ovato : rostro elongato, acuminato, piloso, exceptis naribus parvis convolutisque ; labiis magnis, tumidis, et pilis brevibus consitis : mystacibus plurimis, albis : flocco alborum pilorum seu setarum in mento : genarum ventriculis amplis, disruptis extrinsecè ori, ex supremo labio ad guttur usque protentis ; cavitate retrorsum ad aures pertingente pilosâ : oculis mediocribus : auribus parvis, rotundis, pilose fimbriatis : anterioribus cruribus parvis : pede mediocri, setosis marginato pilis : unguibus brevibus, uncinis, excepto polliculari in orbem figurato vel ad instar humani : posticis cruribus longis ; pedibus magnis validisque, digitis quinque instructis, medio cæteris aliquantulò longiore ; intimo digito brevissimo, attingente tantum metatarsa cæterorum ossa : unguibus omnibus brevibus, acuminatis, modicè incurvis : caudâ longâ, attenuatâ, pilis brevibus sericis coopertâ ; colore suprà sepiaco-brunneo, infrà albo ; obscurâ lineâ transcurrente genas sub oculis.

SYNONYMES.

Cricetodipus Parvus.—Peale, Mamm. of U. S. Exploring Expedition, p. 53.
Perognathus Parvus.—Dr. Le Conte, Proc. Acad. Nat. Sci., Phila.

DESCRIPTION.

Head ovate ; the snout elongate, pointed, and covered with hair, excepting the nostrils, which are small and convolute ; lips large, tumid, and covered with short hairs ; whiskers numerous, white ; a tuft of white hairs or bristles on the chin ; cheek-pouches spacious, opening outside of the mouth, and reaching from the upper lip to the throat ; the cavity extending backwards to the ears, and lined with hair ; eyes medium size ; ears small, round, and fringed with hairs ; fore-legs small, the feet moderate, margined with bristly hairs ; the nails short, curved, excepting that of the thumb, which is orbicular, or resembling the human thumb nail ; hind-legs long ; the feet large and strong, five-toed ; the middle one slightly longer than the rest ; inner toe shortest, reaching only to the end of the metatarsal bones of the others ; all the nails short, pointed, and slightly curved ; tail long, tapering, and clothed with silky hairs. Colour above sepia-brown ; beneath white ; a dark line crosses the cheeks beneath the eyes.

DIMENSIONS.

		Inches.
Length of the head and body, - - - - - -		$1\frac{9}{10}$
" of head, from the nose to the occiput, - - -		$\frac{9}{10}$
" of ears, - - - - - - - - -		$\frac{3}{10}$
" of tail, - - - - - - - - -		$2\frac{3}{10}$
" of fore-leg from the elbow, - - - - -		$\frac{9}{20}$
" of fore-foot, - - - - - - - -		$\frac{3}{10}$
" of tibia, - - - - - - - - -		$\frac{7}{10}$
" of hind-foot, - - - - - - - -		$\frac{8}{10}$
" of metatarsus, - - - - - - -		$\frac{5}{10}$

GENERAL REMARKS.

A single specimen of this singular animal was obtained in Oregon, but no notes were furnished by the person who obtained it. The formation of its hind-legs leaves but little room to doubt that its habits are similar to the jumping mice, *Meriones Labradorius* (RICHARDSON), which are inhabitants of the same region. Its singularly large head, which equals its body in bulk, its ample cheek-pouches, long hind-legs, and long tail, present a general form which is peculiar and altogether very remarkable. On dissection, the stomach was found to contain a pulpy matter, which appeared to be the remains of a bulbous root ; the liver is very large, and consists of five foliaceous lobes ; we were not able to detect any gall-bladder.

The specimen is a female, and presents the rudiments of a fourth molar tooth in each side of the lower jaw, which would eventually have replaced the front ones, already much worn. (PEALE.)

DIDELPHIS BREVICEPS.—Bennett.

SYNONYME.

Didelphis Breviceps.—Bennett, Zool. Proc. for 1833, p. 40.

DESCRIPTION.

Allied to *D. Virginianus ;* much smaller size and darker colour ; ordinary woolly hairs of the body white at base, the apical half, brownish black. Beyond this woolly hair there is an abundance of immensely long bristly white hairs on the upper parts and sides of the body. Head, throat, and under parts of body brownish, the hairs being white, with the tips brown. Lips white ; a broad white dash under the eye, joining the white lips ; a longitudinal brownish stripe extending from the eye towards the tip of the muzzle ; brownish black hairs also surround the eye. On the crown of the head there are long white hairs, interspersed like those of the body, but shorter. Ears black, naked, the apex whitish ; limbs and feet brown-black ; tail with minute bristly hairs, springing from between the scales ; the basal half of the tail apparently blackish, and the apical half whitish.

DIMENSIONS.

	Inches.	Lines.
From point of nose to insertion of tail, - - -	12	6
Tail, - - - - - - - -	11	
Tarsus to end of longest claw, - - - -	2	
Ear to point of nose, - - - - - -	3	
Height of ear posteriorly, - - - - -	1	1
Longest bristly hairs on the back, - - - -	3	

GEOGRAPHICAL DISTRIBUTION.

California. (Bennett.)

DIDELPHIS CALIFORNICA.—Bennett.

SYNONYME.

Didelphis Californica.—Bennett, Zool. Proc. for 1833, p. 40.

DESCRIPTION.

Body above brownish black, intermediate in size between *Didelphis Virginianus* and *D. Breviceps*, head much longer than that of *D. Breviceps*, ears and legs longer, the inner toe on the hind-foot much the longest.

DIMENSIONS.

	Inches.	Lines.
From nose to root of tail,	14	.
Tail,	13	9
From ear to point of nose,	3	11
Tarsus,	2	2
Height of ear posteriorly,	1	9

GENERAL REMARKS.

The description given of the colours of *D. Breviceps* will, in most respects, apply to the present. The long white hairs interspersed throughout the fur, the black line running through the eye, and the black feet and legs, would lead to the supposition that the species were identical.

But there are, notwithstanding, many striking marks of difference. The inner toe of the present species is much larger and far separated from the second, which at first sight seems to be united by a web.

The animal is also more lightly coloured.

Brought by Douglas from that portion of California nearest to Mexico. (Bennett.)

MUS CAROLINENSIS.

CAROLINA MOUSE.

M. Dilute plumbeus, auribus longis et pilosis, cauda corpore longiore.

CHARACTERS.

Tail longer than the body ; ears long and hairy.   Color light plumbeous.

SYNONYMES.

Mus Carolinensis.—Aud. and Bach, Jour. Acad. Nat. Sci., Phila., vol. viii., part ii., p. 306.
  "   "   Le Conte, Proc. Acad. Nat. Sci., Phila., p. —, 1853.

DESCRIPTION.

In size this species is smaller than the house mouse. The upper fore-teeth are slightly grooved. The head is short, the forehead arched, and the nose rather blunt. Eyes small, but prominent ; whiskers longer than the head. The ears are rather long, and have a very conspicuous incurvation of their anterior margins, which are fringed with hairs ; they are thickly clothed on both surfaces with very short hairs. The legs and feet are small and slender, hairy to the nails. The thumb is almost entirely composed of a short convex nail. The tail is long, clothed with short hairs, rounded in the living animal, but square when in a dried state. The fur, which is of moderate length, is thin, soft, and silky.

COLOUR.

The incisors are light yellow, tipped with black ; eyes black ; point of the nose, lips, chin, fore-feet, and nails, white. Whiskers dark brown. There is a narrow fawn-coloured ring around the eyes. Ears, legs, and tail light ashy brown. The fur on the back and sides is from the roots of an uniform light plumbeous colour ; the under surface is scarcely a shade lighter.

DIMENSIONS.

	Inches.	Lines.
Length of head and body, - · - · ·	2	4
" of tail, - · - · · · ·	2	9
Height of ear, - · · · · · ·		4
Length of tarsus, - · · · · · ·		6½

GENERAL REMARKS.

This species exists very sparingly in the maritime districts of South Carolina, and is usually found in low grounds partially inundated. It readily takes to the water, and swims with great facility.

SOREX RICHARDSONII.—Bach.

Richardson's Shrew.

SYNONYMES.

Sorex Parvus.—Rich (non Say), Fauna Boreali Americana, p. 8.
Sorex Richardsonii.—Bachman, Jour. Acad. Nat. Sci., Phila., vol. vii., part -ii.,
    p. 383.

DESCRIPTION.

Ears short, about half the length of the fur, covered by short fine hairs ; muzzle long and slender, the tip slightly lobed ; the whole upper lip bordered with whiskers, reaching to the ears ; the tail square, pointed at tip ; body longer and thicker than that of *S. Forsteri ;* feet slender, partaking, in this respect, of the character of most of the species of this genus ; nails short and slightly hooked.

COLOUR.

The fur, from its roots to near the tip, has a dark bluish grey colour ; from its closeness, however, this colour is not seen till the fur is removed ; the whole upper surface is of a rusty brown colour ; beneath cinereous ; the feet and nails are light brown.

DIMENSIONS.

	Inches.
Length of head and body, - - - - - - -	$2\frac{3}{4}$
" of tail, - - - - - - - -	$1\frac{5}{8}$
" of head, - - - - - - -	$\frac{7}{8}$
" from upper incisors to nostrils, - - - -	$\frac{1}{8}$
" from eye to point of nose, - · · - -	$\frac{7}{16}$

SOREX BREVICAUDUS.—Say.

SHORT-TAILED SHREW.

CHARACTERS.

Blackish plumbeous above, a little lighter beneath ; smaller than S. Dekayi ; tail a little longer.

SYNONYMES.

SOREX BREVICAUDUS.—Say, Long's Expedition, vol. i., p. 164.
    "   "   Godman, vol. i., p. 79, plate 3, fig. 1.
    "   "   Harlan, Fauna, p. 29.
    "   "   Bachman, Jour. Acad. Nat. Sci., Phila., vol. vii., part ii.,
             p. 331.

DESCRIPTION.

The form of this species is more slender than that of DEKAY's shrew, and it appears about one fifth less ; the feet are a little longer and rather large for the size of the animal ; the fur on the back long, nearly double the length of the other species ; the fore-feet are naked ; the hind ones sparsely covered with hair ; the nose is distinctly lobed ; the orifice to the internal ear is large, with two distinct half-divisions ; the tail in the dried specimen appears to be square, sparsely clothed with hair which extends beyond the tip.

COLOUR.

The nose and tail are dark brown ; feet and nails white ; the whole upper surface of a blackish plumbeous colour ; the under surface a little lighter.

DIMENSIONS.

	Inches.
Length from tip of the nose to root of tail, - - -	3¼
" of heel to end of tail, - - - - - -	¾
" of tail, - - - - - - - -	1
" of head, - - - - - - -	¾
Breadth across the head, - - - - - -	½

GENERAL REMARKS.

The teeth of this shrew are white, brightly tinged with chestnut brown on the points, except the third and fourth lateral incisors in the upper jaw, which have merely a brown speck at the tips, and the fifth, which is white ; the posterior upper molar is small, though larger than that of *S. Dekayi ;* the incisors are less curved than those of the latter species ; there is also a striking difference in the head, that of the present species being considerably shorter, the skull more depressed and much narrower, appearing about one fourth less than that of DEKAY's shrew.

From the number and appearance of its teeth, the specimen was evidently an old animal.

PSEUDOSTOMA BULBIVORUM.—Rich.

SYNONYMES.

Diplostoma? Bulbivorum.—Rich, Fauna Boreali Americana, p. 206.
Diplostoma (Geomys) Bulbivorus.—Rich, Zoology of Beechey's Voyage, p. 13.
Geomys Bulbivorus.—Dr. Le Conte, Proc. Acad. Nat. Sci., Phila., 1852, p. 162.

DESCRIPTION.

Body like that of a great mole ; furnished with cheek-pouches, each pouch has a semi-cup-shaped cavity when distended ; whiskers very short, eyes small. The *auditory openings* are moderately large, but there are no external ears. Tail short, round, and tapering, with an obtuse tip, and thinly clothed with hair. The legs are short, and are covered down to the wrist and ankle joints with fur similar to that of the body ; there are five toes on each foot ; the hind nails are short, conical, obtuse, and more or less excavated underneath. The nail of the fourth toe is more spoon-shaped than the others.

COLOUR.

Incisors yellowish ; on the dorsal aspect the fur has a colour intermediate between chestnut and yellowish brown, darker on the crown of the head than elsewhere ; on the belly the brown is mixed with a considerable portion of grey. The lips, the lower jaw, the lining of the pouches, and a narrow space around the arms, are covered with white fur.

Close to the upper part of each side of the mouth there is a rhomboidal mark, which is clothed with hair of a liver brown colour.

The *hind-feet* are covered above with whitish hairs.

DIMENSIONS.

	Inches.	Lines.
Length of head and body,	11	
" of head,	3	
Breadth of head behind the eyes, when the pouches are distended,	3	6
Length of tail,	2	6
" of upper incisors (the exposed portion),		6
" of lower incisors,		9

GENERAL REMARKS.

We have altered in arrangement, and abridged, Sir JOHN RICHARDSON'S description of this pouched rat, which, in some particulars, has so very great a resemblance to *P. bursarius*, as to have made us hesitate to place it in our work.

DIPODOMYS AGILIS.—Gambel.

CHARACTERS.

Tail brownish, with an indistinct whitish vitta on each side ; outer third to tip nearly uniform pale brown.

SYNONYMES.

Dipodomys Agilis.—Gambel, Proc. Acad. Nat. Sci., Phila., vol. iv., p. 77.
    "   "   Dr. Le Conte,   "   " vol. vi., p. 224.

DESCRIPTION.

In the upper jaw the incisors are divided by a longitudinal furrow ; head elongated, tapering from the ears to a sharp point ; ears nearly round, sparsely hairy ; eyes large ; a large pouch on each side of the head opening externally on the cheeks. Both hind and fore-feet with four toes and a rudiment of a fifth. Hind-legs very long ; tail strong, very slender, covered with hair, and ending in a penicillated tuft.

Two incisors and eight molars in both upper and lower jaws.

COLOUR.

Above, yellowish brown mixed with dusky ; beneath, pure white, extending half way up the sides ; eyes dark brown.

DIMENSIONS.

	Inches.
Total length, including the tail, - - - -	$10\frac{1}{2}$
Length of tail, - - - - - - - -	$6\frac{1}{4}$

HABITS.

This beautiful Jerboa-like animal is abundant in the vineyards and cultivated fields of the Pueblo de los Angeles, Upper California. Like the other pouched animals it forms extensive burrows, traversing the fields in different directions, and is only to be dislodged during the process of irrigation. It leaps with surprising agility, sometimes the distance of ten feet or more at a spring, and is difficult to capture. (Gambel.)

DIPODOMYS HEERMANNI.—Le Conte.

SYNONYME.

Dipodomys Heermanni.—Le Conte, Proc. Acad. Nat. Sci., Phila., vol. vi., p. 224.

DESCRIPTION.

Tail shorter than the body ; hairs on the outer third very long ; ears moderately small ; antitragus obsolete.

COLOUR.

Tail brown, becoming black towards the extremity, with a broad white vitta on each side ; tip pure black.

GENERAL REMARKS.

This species was procured in the Sierra Nevada, by Dr. Heermann. The specimen was not quite full grown. The above description, &c., we take from Dr. Le Conte's remarks in the Proc. Acad. Nat. Sciences, Philadelphia, cited above.

PEROGNATHUS FASCIATUS.—Wied.

P. Supra e flavescente cinereus, subtus albus, strigâ laterali pallide flavus.

SYNONYMES.

Perognathus Fasciatus.—Wied, Nova Act., Leopold Car. Acad., 19, 369, tab. 34.

"   "   Wagner, Schreber's Saügthiere, Suppl. 3, 612.

"   "   Schintz, Syn. Mam., 2, 259.

DESCRIPTION.

The upper surface is brownish-grey ; the hairs at the roots olive-grey, at the tip yellowish and blackish, whence the animal appears speckled with blackish and fulvous, or is somewhat striped.

The sides of the head, the region round the eyes, and upper margin of the ears, are of a more dull reddish-yellow. The under surface is pure white, which is separated from the colour of the back by a yellowish-red, or rust-red stripe extending from the nose along the whole side to the hind-legs, and down to the heel. The nose and lips appear flesh coloured through the whitish hair ; the same is the case with the legs below the tibiæ. The tail is reddish-grey ; more greyish above, and more whitish beneath.

DIMENSIONS.

	Inches.	Lines
Entire length, - - - - - - -	4	4½
Length of tail (including hair), - - - -	2	1
" of fore-foot, - - - - - -		3½
" of hind-foot, - - - - - -		8

GEOGRAPHICAL DISTRIBUTION.

This species was procured in the territories west of the State of Missouri.

SCIURUS CLARKII.—Smith.

CLARK'S SQUIRREL.

SYNONYME.

Sciurus Clarkii, Clark's Squirrel.—Griffiths, Cuvier, vol. iii., p. 189.

DESCRIPTION.

Back, upper parts of the head and neck, cheeks and tail, of a delicate silver grey colour ; the shoulders, flanks, belly, and posterior extremities, both within and without, are white with a slight ochreous tint ; on the sides of the nose and the fore-legs this tint deepens in intensity ; the head is rather flattened and thick, the ears small and round ; eyes black, and situated on the sides of the head very far distant from each other, leaving a wide expanse of forehead. The nostrils are semilunar in shape ; the upper lip is cleft, and there is a black spot on the chin.

The tail, which is flat and spreading, is very beautiful, not so full near its interior as towards the middle, and again diminishing in breadth until it terminates in a point.

GENERAL REMARKS.

We are greatly inclined to consider this squirrel as identical with *Sciurus Fossor*, of PEALE, which we have figured and described. Should other specimens of this species not be found and more positively determined, it would perhaps be better to retain the name of SCIURUS CLARKII, and give *S. Fossor* as a synonyme.

SCIURUS ANNULATUS.—Smith.

SYNONYME.

Sciurus Annulatus, Lewis's Squirrel.—Griffiths, Cuvier, vol. iii., p. 190.

DESCRIPTION.

Has the upper part of the head, neck, shoulders, fore arms, to the articulation of the arm, back, flank, the posterior moiety of the thighs and a band round the belly, of ochrey-grey colour ; all the under parts, the inside of the limbs, and the paws are pure ochrey ; the ears are small, round, and far back ; the eyes are black, and surrounded with the same colour as the back ; the nostrils open at the extremity of the muzzle, forming a denuded black snout ; the upper lip is white and the whiskers very long. The tail is very beautiful, extremely thick and bushy, cylindrical, and annulated with seven black and six white bands, with the termination black.

"This appears to be the *S. Annulatus* described by Desmarest, Encyclop. Method., article Mammalogie. His specific characters are : Fur of a bright greenish grey above, with lateral white bands, white underneath, tail longer than the body, round, annulated black and white." (Le Conte.)

GENERAL REMARKS.

This animal was, as well as *Sciurus Clarkii*, brought from the north-west by Lewis and Clark, on their return from their celebrated journey across our continent.

The specimens were deposited in Peale's Museum, in Philadelphia, and were, it is supposed, burnt up when the remains of that collection were destroyed by fire.

Unless the peculiar *annulated* tail was the result of twisting that member when the animal was skinned, it is difficult to suppose this to have been a true squirrel. We do not know, however, any spermophile that will agree with the description of it.

We have above given descriptions of some quadrupeds which we have not ourselves had an opportunity of examining—the result of the observations of other zoologists—but are not at present able to state positively that *all* of them are founded on good species.

We add some names of animals that have been given by authors as belonging to our Fauna, but which we have not been willing to introduce as such into our work, and which may, we think, be safely omitted in future lists.

Sorex Cinereus.—Bach. Young of *S. Carolinensis.*

Ursus Arctos.—Rich. A doubtful species.

Sciurus Texianus.—Bach. Grey variety of *S. Capistratus*, without white ears.

   "   *Occidentalis.*—Bach. Variety of *S. Auduboni.*

Arvicola Nuttalli.—Harlan. Young *Mus leucopus.*

Mus Virginicus.—Gmel. Probably an albino of *Mus leucopus.*

Lepus Campestris.—Bach. This appears to be identical with *S. Townsendii*, and we should have given the latter name as the synonyme.

Lipuria Hudsonica.—This is supposed to have been a distorted or mutilated skin. There is no animal to correspond with the description of it.

Lepus Longicaudatus.—This is an African species from the Cape of Good Hope.

Felis Occidentalis.—Probably *Lynx rufus.*

   "   *Fasciata.*   "

Lutra Californica.—Grey, supposed to be *L. Canadensis.*

Felis Discolor.—*Felis Concolor.*

Condylura Macroura.—*C. Cristata.*

Mus Agrarius.—Godman. *Mus leucopus.*

Spermophilus Beecheyi.—Rich. *S. Douglassii.*

Sorex Talpoides.—Gapper. Probably *S. Carolinensis.*

Ixalus Probaton.—A hybrid.—Not American.

Cervus Arctica.—Rich. Requires further examination.

Lepus Mexicanus.—*L. Nigricaudatus.*

Sciurus Aurogaster.—*S. Ferruginiventer.*

   "   *Californicus.*—*S. Nigrescens.*

Sorex Canadensis.—*Scalops Aquaticus.*

Saccomys Anthopilus.—South American.

Spermophilus Pealei.—Not American.

INDEX.